Optical Fibers and RF:
A Natural Combination

Optical Fibers and RF: A Natural Combination

Malcolm Romeiser

NOBLE
PUBLISHING

Noble Publishing Corporation
Atlanta, GA

Library of Congress Cataloging-in-Publication Data

Romeiser, Malcolm
   Optical fibers and RF : a natural combination / Malcolm Romeiser.
   p. cm.
   Includes index.
   ISBN 1-884932-34-7
   1. Optical communications.   2. Optical fibers.   I. Title.

TK5103.592.F52R65 2003
621.382'75—dc21

                                        2003051041

NOBLE
PUBLISHING

Copyright 2004 by Noble Publishing corporation.

Printed in the United States of America

ISBN 1-884932-45-2

Contents

Preface

I will use the first person in writing this preface because it will contain many of my thoughts on this technology, how it has grown, and how it impacts society. The remainder of this book is in the third person, which is the usual way to present technical material.

This book was written with three objectives in mind: first, and foremost, as a teaching text; second, to present updated information on optical fiber technology in a practical context useful to students or professionals; and third, to provide a platform for presenting different insights and analyses on certain optical fiber system design issues. The ability to achieve the first comes from my experiences through the last decade while teaching undergraduates at a university and as a consultant. The third comes from my earlier years in industry, where I helped research, design and develop transmission systems that used radio, wire, coax, and optical fibers. From this broad base of practical and theoretical knowledge I gained an ability to analyze and simplify telecommunications transmission systems issues. The second objective is a blend of the other two. I have noticed that very useful information and insights are lost when too much technical detail is presented. Having experienced the growth of the technology and also having developed an understanding of how it can best be taught, I feel compelled to share these experiences.

The way I approach optical fiber technology in this book probably owes more to my understanding of today's undergraduates than to my industrial experiences. There was a great temptation in writing this book to do what most authors do: present many equations without derivation or explanation so as to cover more details in the allotted space. Unlike earlier generations, however, present-day engineering students are not comfortable with this approach. For whatever reasons, they find it hard to "go learn it on your own". Some amount of forced independent analysis is unavoidable, but hopefully I have kept it to a minimum. Like my college students, the reader is expected to have a working knowledge of differential calculus and basic physics.

The book's title was inspired by a recognition that radio technology and optical fiber technology have much in common. I believe optical fibers should be viewed as the latest technology, in a century long progression of technologies, that allow the efficient transmission of information. One obvious link between the two is that transmission is achieved using the same three-step sequence: carrier modulation at a transmitter, propagation through a media, and demodulation at a receiver. The carrier frequencies and devices differ between technologies but many of the transmission techniques are similar. Since optical systems transmit radio frequencies, there is also ongoing synergy between RF and optical fiber research and development.

Another inspiration for the title came from a study of the history of radio development. Radio frequency technology began about 100 years before the beginnings of optical fiber technology. Optical fiber transmission depends on advances made by early radio researchers. One example is the fundamental research on electromagnetic (EM) wave propagation performed by C. Maxwell and W. T. Kelvin in the late 19th Century. This proved essential in achieving transmission of information first over long and short wave radio, then wire pairs, coaxial cables, line-of-sight microwave radio, and finally optical fibers. Each of these media in turn provided increasing amounts of bandwidth, increased information capacity, and greater transmission distances. Many other connections from radio system development to optical fiber system development exist. Some examples are: the modulation and demodulation of signals on carriers, the filtering and separating of these signals at high frequencies, and the frequency division multiplexing of many independent signals.

Optical fiber technology has grown rapidly in the period from1970 to the early 2000's. Current popular opinion holds that the pace of adoption of optical fiber technology has been unprecedented. This is actually not true when compared to the historical development of other innovative technologies like radio, transistors, integrated circuits, plastics, etc. Probably the most surprising parallel between RF and optical fiber development is that both required about an equal length of time to progress from initial theoretical studies to significant commercial use, a period of about 35 years.

At the time of discovery, both technologies were too impractical to draw much attention. At the turn of the 20th Century G. Marconi, the generally acknowledged father of radio, had great difficulty interesting anyone in its commercial possibilities. As for optical fibers, this author can personally attest to an early lack of commercial interest. I organized a conference session on optical fibers in the early 1970s that I was convinced would be very well attended. After all, this new technology was going to be the greatest thing since sliced bread. Only about 10 people showed up. I realized that even the most promising technology is of little interest if it has no immediate commercial value.

For both technologies, rapid commercialization ensued once reasonably priced components became available. In both cases this took about ten years from the time of the seminal work. The fundamental theories underlying radio transmission were developed in the late 1800s. The enabling technologies were developed soon after the turn of the century. By the late 1930's there was widespread use of radio for communications. The fundamental optical fiber work began in the mid-1960s. The enabling technologies were first available commercially in the 1970s. By 2000 the technology was in widespread use.

More detailed comparisons can be drawn at the component level. In the very early years, radio transmission used telegraph signaling and spark-gap transmitters. The RF frequencies were low (60 kHz) which resulted in high atmospheric attenuation and limited transmission distances. The transmitted RF spectra were broad and the detectors used intensity sensitive metal filings that changed resistance according to the received EM energy. The antennae were nondirective which meant the media was used very inefficiently. Compare this to early work with optical fibers. The only light sources available were at 800 nm. These wavelengths are in the lower, high attenuation range of useable wavelengths for glass fibers. As a result, transmission distances were limited. The light sources had broad spectra, and the detectors were sensitive only to intensity variations. As an aside, the detectors used in modern systems still use intensity detection. The fibers were large-core, multimode, which resulted in inefficient media usage.

For radio, the most important component advance was probably the invention of the thermionic valve (vacuum tube triode) in 1908. This allowed signal amplification and more advanced modulation/detection methods. A comparable advance with optical fiber technology was the development of the room temperature, 850 nm semiconductor laser and the etched well LED in the early 1970s.

When a baseband signal is modulated on an RF carrier the resulting spectra can be wide or narrow. The type of modulation determines the spectral width, as well as the sensitivity of the signal to noise, distortion, and interferences. If the spectra are narrow the transmission media can be used more efficiently but, in general, the system costs will be higher. The spark-gap/telegraph transmission used for the first radio systems was inefficient and was replaced by amplitude-modulated high frequency carriers once vacuum tubes were developed. Another important advance was the development of single-sideband suppressed carrier (SSBSC) transmission in the 1920s. SSBSC transmission basically allowed the RF energy to be concentrated in the signal and not in a carrier, thus improving transmission quality and efficiency. The ability to generate high frequency, coherent carriers also allowed accurate, predictable and reliable signal filtering to be achieved. Also, longer transmission distances and greater total bandwidths were possible through the use of higher frequencies.

With SSBSC, independent (modulated) carriers are transmitted in closely packed adjacent bands. The term applied to this approach is frequency division multiplexing (FDM). With optical fibers the comparable technique is called wavelength division multiplexing (WDM). WDM became realizable and economically attractive only with the availability of longer wavelength sources and detectors. At longer wavelengths glass fibers have less attenuation and distortion. Widespread use of WDM began in the early 1990s, about 15 years after the beginning of fiber system commercialization. The more advanced WDM systems use nearly coherent (single frequency) laser sources and intensity modulation and detection. This modulation approach produces a double sideband amplitude modulated signal like that used by radio before SSBSC. At some future date optical fiber systems will undoubtedly adopt more efficient types of modulation like SSBSC.

Why dwell on these comparisons? As I mentioned, my view of optical fiber technology is that it is just another step in a long progression of telecommunications technologies. I began my telecommunications career working on analog microwave radio. I have had the advantage of working on many technologies since then: analog satellite, digital wire pair, digital coax, digital microwave, analog cellular radio, digital optical fiber systems. For each of these technologies the media and required hardware are unique, but the underlying communications principles are the same. In writing this book I have attempted to concentrate on what makes optical fiber transmission unique, while at the same time, trying to position the technology as just another step on this evolutionary path.

Optical fiber technology is both broad and deep. Understanding the technology thoroughly requires that knowledge be obtained from many disciplines: physics, engineering, optics, EM wave propagation, semiconductor theory, electronic circuits, and communication theory. Because the intent in this book is to present the technology in an overview, it has been necessary to leave out many details. Unfortunately some important topics are treated very briefly. The reader is encouraged to seek additional information. To that end a limited list of references has been appended.

This book begins with a discussion of underlying physical concepts important to optical fiber systems. Chapter 2 discusses the construction and propagation of light in the two basic types of cylindrical glass fibers, multimode and single-mode. Non-cylindrical optical waveguides are used in optical integrated circuits and some optical sensors. This use is mentioned in Chapter 6, without a derivation of the waveguide characteristics. Chapter 3 presents the performance characteristics of cylindrical fibers. Their ability to transmit information is dictated by two parameters, dispersion and attenuation. Both play a role in determining the available system bandwidth and the maximum transmission distance. The devices used for generating and detecting light signals are discussed in Chapters 4 and 5. An important characteristic of the light, coherence, is covered in Chapter 4. Radio signals generated to carry modulation or help in the detection process at a receiver are very coherent. This allows the RF bands to be used with a high degree of spectral efficiency. Optical carrier sources still have significant coherence limitations. Chapter 6 discusses the optical components needed to assemble and test an end-to-end system. These components provide many important functions: connecting fibers, controlling levels and reflections of light energy, separating and combining signals, switching, and modulating. Chapters 7 and 8 combine the concepts developed in Chapters 1–6 Chapter 7 presents two basic system designs, a short distance analog video link and a long distance digital link. Chapter 8 deals with the current trends in optical fiber use and where the technology might be headed in the future. The spread of Internet packet switching is probably the most important factor in how this technology will be used in the telecommunications arena.

Finally, I would like to acknowledge the encouragement received from my family and friends. Especially I appreciated the support and patience of my wife, Sonja.

1

From RF to Optical

1.1 Introduction

Through most of the 20th Century the basic media used to transmit radio frequency (RF) energy were: wire pairs or conductors, coaxial cable, waveguides, and the atmosphere. The first two require electron flow in metallic conductors to direct the RF propagation. The last three rely on propagating electromagnetic fields, guided either by a confining structure or directive antennas. These media have provided efficient transmission capability since the early days of radio. The term *media* is generic, referring in general to the substance supporting the wave's propagation.

"Radio" is defined in the dictionary as the use of electromagnetic waves in the frequency range from 10 kilohertz (kilocycles/sec) to 300 Gigahertz to transmit electric signals without the use of wires. The upper range of modern radio frequency (RF) applications now extends beyond 40 Gigahertz, almost reaching the limit of the radio definition. Achievements in the field of electronic technology are one big reason for this progress. A growing need for mobile communications makes it difficult to know if the technology is pushing or the market is pulling. Beyond 300 Ghz, where frequencies approach the near infrared, another technology is also helping to spur growth in RF applications — optical fibers. Optical fibers are small diameter cylindrical glass filaments that transmit electromagnetic (EM) waves at frequencies near the visible portion of the spectrum.

Light differs from radio because it has both the properties of particles, called photons, and the properties of EM waves. The term "photonics", which is commonly applied to optical fibers, only partially describes the technology. Both particle physics and EM wave theory are needed to completely understand how optical fiber systems work. Particles are characterized by their energy, momentum, and the fact that they can be individually, physically identified. EM waves are characterized by frequency, wavelength, polarization, and velocity.

This chapter initially presents the basic physical parameters that describe wave and signal propagation in Section 1–2. The different media that support propagation have different characteristics and information capacities. The one of most concern for optical fiber systems is the index of refraction. *Refraction* refers to the bending of a propagating wave at a boundary. The refractive index of a media also determines the speed of propagation of the wave in the media. Section 1–3 discusses the three basic elements of any optical fiber system in very broad terms. Subsequent chapters give much more detail about the three: Chapters 2–3 on fibers, Chapter 4 on optical sources, Chapter 5 on optical detectors and receivers, Chapter 6 on components, Chapter 7 on optical system design, and Chapter 8 on future directions for the technology. Section 1–4 gives a brief summary of optical fiber history. Section 1–5 lists units and constants used throughout the book.

1.2   Parameters

1.2.1   *Frequency/Wavelength/Bandwidth*

Figure 1-1 compares the approximate ranges of transmission frequencies used on five basic media: wire pair conductors, coaxial cables, metallic waveguides, the atmosphere, and optical fibers. Optical fibers were added to the media mix beginning about 1980. Optical fibers use specially constructed glass, plastic, or ceramic cylindrical waveguides to contain the propagating electromagnetic fields. Notice in Fig.1-1 that optical fibers are very high on the frequency scale, so high in fact that until very recently it was customary to describe optical energy in terms of wavelengths (λ) instead of frequency (f). As a point of reference, the visible spectrum runs from about 0.4×10^{-6} meters (blue) to 0.6×10^{-6} meters (red). If we used frequencies to describe this energy we would be using units of terahertz.

The most commonly used optical fibers are constructed of silica, which is silicon dioxide (SiO_2). Silica is common window glass, but for use in fibers it has been purified and doped to reduce energy loss. The doping can be accurately controlled so that the characteristics of the waveguide can be customized for specific applications. The optimum performance wavelength range for silica fibers is from 0.8×10^{-6} meters to 1.6×10^{-6} meters. The MKSC system of units (meter-kilogram-second-coulomb) will be used in this text, and is listed at the end of this chapter.

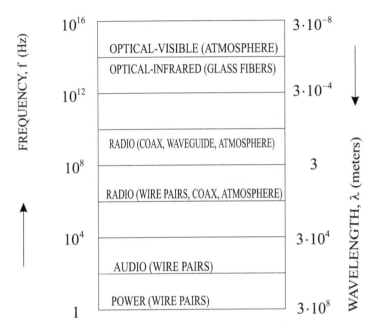

Figure 1-1 *Electromagnetic spectrum.*

Optical fiber technology requires the use of very small and very large numbers. A wavelength value is often given in nanometers, which is a billionth of a meter. This same wave might also be valued using a frequency in the terahertz range, which is a thousand-billion hertz. Scientific notation, based on powers of 10, has to be used to describe these very small and very large numbers. Table 1-1 lists important scientific notation prefixes, symbols, and their corresponding factor of 10 multipliers. For example, a length of 10^{-6} meters can be described as 1 μm or 1 micron. Also, 1 μm is equivalent to 1000 nm, because the factor $1\mu m = 10^{-6}$ m $= 1000 \times 10^{-9}$ m $=$ 1000nm. Physicists often use a unit called an Angstrom (Å) to describe very small wavelengths. An Angstrom equals 0.1 nanometers (nm).

One of the obvious advantages to operating at higher frequencies is the increase in bandwidth available for transmitting information. Because of component and system design limitations, only a small percentage of the bandwidth around the center frequency of a communication channel is used for information. For example, commercial AM radio uses about 1 % of the bandwidth around the carrier. Very high carrier frequencies have greater information bandwidth available, which is a fundamental advantage with

Table 1-1 *Scientific notation.*

Prefix	Symbol	x10 Multiplier
tera	T	10^{12}
giga	G	10^{9}
mega	M	10^{6}
kilo	k	10^{3}
centi	c	10^{-2}
milli	m	10^{-3}
micro	μ	10^{-6}
nano	n	10^{-9}
pico	p	10^{-12}
femto	f	10^{-15}

optical fiber systems. Modulated laser bandwidths for optical fiber transmission currently are about 100 GHz wide, which is about 0.1 % of the carrier frequency. Whether optical fiber systems reach the bandwidth efficiencies of broadcast radio is problematic; the tremendous information capacity this implies is probably not needed on a single carrier in the foreseeable future.

1.2.2 *Power*

Communication systems also have a wide range of power values. A laser transmitter array might be radiating one hundred milliwatts into a fiber. A detector at the end of the fiber might be receiving one ten-thousandths of a milliwatt, a difference of 10^6. Wide ranges of relative powers like this are compared using decibels. A decibel is a ratio of two powers or two voltages:

$$dB = 10\log\left(\frac{P_1}{P_2}\right) \qquad 1.1$$

$$dB = 20\log\left(\frac{V_1}{V_2}\right)$$

The two powers could be the input and output powers of the line above, for example. Instead of dealing with numbers differing by a million they would be compared using a decibel value of 60. Often a decibel value will be given relative to a milliwatt. The laser transmitter above would be radiating + 20 dBm. In equation 1.1 P_1 would be 100 and P_2 would be 1 (mW). The signal at the detector would be 0.0001 milliwatts, or −40 dBm. In this case using equation 1–1, P_1 would be 0.0001 and P_2 again would be 1 (mW). The decibel scale is also very useful in describing the response of filters, analyzing noise, and calculating amplifier gain.

1.2.3 *Velocity*

In Figure 1–1 the mathematical relationship between the frequency scale on the right and the wavelength scale on the left is given by:

$$\lambda = v \:/\: f \text{ where} \qquad\qquad 1.2$$

v = velocity of energy propagation in the media

Every media has unique characteristics that determine the velocity. Only in a vacuum, or free space, is the velocity equal to the speed of light, c:

$$c = 3 \times 10^8 \text{ m/sec} \qquad\qquad 1.3$$

Further, in free space c is constant with frequency. In every media other than a vacuum the velocity will, unfortunately, exhibit some degree of frequency dependence. Since most RF signals have multiple frequency components, generally because of modulation sidebands on a carrier, this dependency of velocity on frequency/wavelength can present significant problems. One of the terms used to describe this phenomenon is called *dispersion*.

The velocity of light in water, glass, or media other than free space is slower and given by:

$$v = c/n \quad \text{where} \qquad\qquad 1.4$$

n = *index of refraction* of media

Table 1-2 lists the indices of refraction for common materials used to fabricate the different components of an optical fiber system. Most of these materials are dielectrics. Dielectric materials do not conduct electricity unless specifically doped with current carriers. All semiconductors used in electronics and optical fiber systems have to be doped with either extra electrons or the lack of extra electrons (called holes). The index of refraction of a material is usually stated relative to the free space value of 1. In circular optical fibers the center section, or core, has a higher index of refraction than the outer section, or cladding. In circularly symmetric glass fibers the core's index will be about 1.5 and the cladding's slightly lower.

Electromagnetic (EM) optical waves are slowed in a medium with an index of refraction greater than 1. As a result, when two substances of differing index of refraction are brought together, the optical wave is refracted or bent in the second substance. The wave is either speeded up or slowed relative to the first. The bending occurs either toward a normal to the surface (Index 1< Index 2) or away from the normal (Index 1> Index 2). In an optical fiber the central core has a higher index of refraction than the outer cladding, so light impinging on the core-cladding interface is bent back to

Table 1-2 *Indices of Refraction.*

Material	Index of Refraction n
Air	1.0
Water	1.33
Fused Silica (Quartz)	1.46
Glass	1.5
Polystyrene	1.59
Diamond	2.42
Gallium Arsenide	3.35
Silicon	3.5
Aluminum Gallium Arsenide	3.6
Germanium	4

the center and guided down the fiber. This phenomenon is called *total internal reflection*. The behavior of the optical EM wave in a substance is also dependent on how the refractive index might change with temperature, pressure, time, interfaces, contaminates, etc.

EM wave theory is used to analyze the propagation of light in the optical fiber medium. The other property of light needed to analyze optical fiber systems is the behavior of the photon. Photons are particles of energy. Photon behavior is used to describe the performance of lasers and light detectors. The energy of a single photon of frequency f is:

$$W \text{ (in Joules)} = h \times f = h \times c/\lambda \quad \text{where} \qquad 1.5$$

$$h = \text{Planck's constant} = 6.626 \times 10^{-34} \text{ J} \times \text{sec}$$

The energy of a light wave can be described either in terms of the power in its EM wave using Maxwell's theory, or as the total energy carried by its photons. A light beam will contain an enormous number of photons. Consider for example a light wave of 0.8 μm having known power of 1 microwatt impinging on a detector for 1 second. From equation 1.5, the energy of a single photon is 2.48×10^{-19} J. Since power is the time rate of delivery of energy, 1 microwatt over 1 second gives a total energy of 1 microJoule. Dividing the 1 microJoule by the energy of a single photon gives the number of photons, 4.03×10^{12}.

1.3 An Optical System

An optical system can be used for communications, sensing, or controlling other systems. In almost all applications the system has the three basic components shown in Fig. 1-2: an optical source; the media, which is generally an optical fiber; and an optical detector. The source often is modulated with an electrical signal, hence the source is also shown as an *electrical to optical converter (E/O)*. Similarly, the detector will generally have to convert the optical signal to an electrical signal, hence is also shown as an *optical to electrical converter (O/E)*.

There are two basic types of optical sources, lasers and leds. The term *laser* stands for l̲ight a̲mplification by the s̲timulated e̲mission of r̲adiation. The term *led* stands for l̲ight e̲mitting d̲iode. For optical fiber applications these devices are usually semiconductors, which are small in size, consume little power, are easier to couple to fibers and integrate, and are low in cost. The disadvantage of using semiconductors is their limited light power outputs. Lasers are constructed so that at high electrical drive levels they increase their output through an internal oscillation. This stimulated emission allows high output powers and narrow source bandwidth. Leds and lasers at low drive levels emit light through spontaneous emission, which is basically a random process and broadband.

Detectors convert the energy in photons to electrical current, which then allows the recovery of the original signal. Optical fiber detectors are also usually made of semiconductor materials: Silicon, Germanium, or III-V compounds such as InGaAs. The choice depends on the wavelengths to be detected and the desired performance requirements. It is possible to achieve internal gain in some detectors, called avalanche photo-detectors, through the use of high voltages.

The electrical signals that modulate the carrier can be either in an analog or digital format. They are often referred to as a *baseband* input signal. An analog signal reproduces the originating signal as accurately as possible. As a result, the optical system needs to allow as little distortion and added noise as possible. A digital signal uses pulses, generally all of the same level and shape, to represent the original signal. This requires more complicated coding and modulation/demodulation, but provides many noise and cost advantages. It seems illogical that lower costs can be achieved with digital signaling because of its added complexity. The reason lies in the tremendous progress made in improving the capabilities and lowering the costs of the integrated electronics used for digital signal processing (DSP)

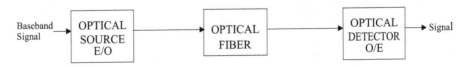

Figure 1-2 *Optical fiber system.*

and transmission. Digital signaling is particularly well suited for the devices used in optical sources and detectors.

For the sake of simplicity, some very important parts have been left out of the system block diagram in Figure 1-2. The electrical signal, in addition to being shaped and coded, will probably be part of a *time-division multiplexed (TDM)* bit stream. The source might be internally or externally modulated. The source has to be carefully coupled to the fiber, and might be combined with other optical signals using *wavelength division multiplexing (WDM)*. Optical connectors might be used to help bring all the fibers together in a patch panel. The fiber will probably be part of a cable that contains other fibers and, perhaps, wire pairs or coax to provide operational flexibility. The fiber might be large core multimode or small core single-mode. A long fiber connection will require splices at intermediate points. At the receiver components are required in reverse order to restore the electrical signal. All of these elements will be discussed in more detail in subsequent chapters. A short single fiber optical system used for sensing will have only a few of these extra parts, while a long telecommunications system will have all of them.

1.4   Optical Fiber History

The rate of penetration of new technologies into our lives has been increasing ever since the dawn of the industrial age. Using one definition of when a new technology gains widespread use, the telephone took about 70 years. The auto took a similar length of time. Personal computers, the Internet, and optical fibers have taken only about half that time. The *Information Age* is truly upon us because information, in and of itself, now is a commodity of value. At the beginning of the 21st Century essentially all long distance communications is carried on optical fibers.

Optical systems have a long history because man has always used vision as a means of communicating. The Greeks received word of the fall of Troy using a fire telegraph; beacons were started and stopped much like a computer's modem communicates. They used an array of torches to convey an element in a matrix of symbols. In the 17th and 18th centuries much was learned about the nature of light through the efforts of F. Grimaldi, I. Newton, R. Boyle, C. Huygens, A-J Fresnel, and W, Herschel. In 1887 H. Hertz demonstrated the particle or photon nature of light. In 1905 A. Einstein explained the photoelectric effect. In 1923 by O.V. Lossew observed the LED effect. In 1880 Alexander Graham Bell invented and demonstrated the photophone, a device that used voice-modulated sunlight. The receiver used a photoconducting selenium cell to demodulate the voice.

Also in the 19th Century, S. Morse developed a wire-based telegraph code that is still in use today. G. Marconi first demonstrated radio transmission in the late 1800s. Electric and magnetic field theory was unified by C.J.Maxwell, which was the major advance needed to quantify EM propagation.

In the late nineteenth century a heliograph system was operational in Arizona and New Mexico. It was used effectively in the Apache Indian wars. Heliography uses modulated sunlight, reflected and directed off mirrors, to communicate between mountaintops. The total system was 809 miles long, and connected many cavalry posts and settlements. The world heliograph distance record was set with a flash of 125 miles. The atmosphere was very clear in the southwest at that time. Average communication speed, in the digital terms of today, was 10 words per minute, which equates to about 7 bits/sec. Compare this result to a modern optical fiber carrying 40 gigabits/second.

In 1854 John Tyndall demonstrated the phenomenon of total internal reflection that is the basis for optical fiber transmission. A patent on guided optical transmission over a glass-type waveguide was obtained by the AT&T Corp. in 1934. The modern advances that have lead to today's widespread use of optical fibers began in the 1960s, but this progress would not be possible without the basic work of the earlier researchers.

In 1966, C. Kao published a paper that proposed using optical fibers for a telecommunications transmission system. A room-temperature semiconductor laser and low loss glass fibers appeared to be realizable, which made a total transmission system feasible. Figure 1-3 illustrates the progress through the years in achieving pure, low loss glass. The glass used by the Egyptians probably only allowed a determination of whether it was day or night outside. Today's optical fibers use glass with much lower impurity concentrations.

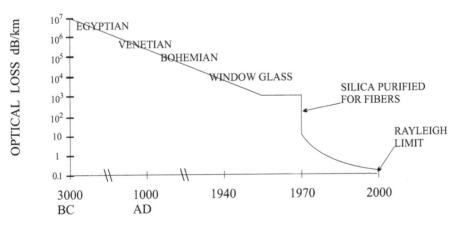

Figure 1-3 *Optical loss in glass.*

By 1970 silica glass waveguides were fabricated with losses less than 20 dB/km. This is the generally accepted point in time where optical fiber transmission systems were judged to be achievable. By the early 80's losses were reduced to the absolutely lowest level for glass waveguides, about 0.16 dB/km at a wavelength of1550 nm. This is called the *Rayleigh scattering limit*. Now it is possible to make very pure (low loss) fibers that have precisely controlled core refractive index shapes. This allows great flexibility in meeting the demands of the modern world for fibers with different characteristics.

1.5   Units and Constants

Table 1-3 lists the basic MKSC (meter-kilogram-second-coulomb) international system units used in this book. Optical fibers have become an important international technology because of the globalization of telecommunications. The development and use of technical standards, derived by international groups, are now very important to the growth of the technology. The reader is referred to other sources for a more detailed discussion of these units, as well as the equivalency between English and metric units.

Table 1-4 lists the primary constants used in this text. These are used in numerous equations found throughout the book.

Table 1-3 *Units.*

Unit	Measures	Symbol
ampere	current	A
coulomb	charge	C
decibel	x10 multiplier	dB
degree Kelvin	temperature	K°
degree Celsius	temperature	C°
degree Fahrenheit	temperature	F°
farad	capacitance	F
hertz	frequency	Hz
joule	energy	J
kilogram	mass	kg
meter	distance	m
newton	force	N
ohm	resistance	Ω
second	time	s
volt	voltage	V
watt	power	W

Table 1-4 *Constants.*

Description	Symbol	Value
Boltzman	k	$1.38(10)^{-23}$ J/K°
electron charge	-e	$-1.6(10)^{-19}$ C
electron volt	ev	$1.6(10)^{-19}$ J
Planck	h	$6.626(10)^{-34}$ J/s
velocity of light in a vacuum	c	$3(10)^8$ m/s

2

Optical Fiber Characteristics

2.1 Introduction

Microwave frequencies are propagated on coaxial cables, inside metallic waveguides, or through-the-air. Optical frequencies are propagated on
dielectric (non-metallic) fibers, waveguides embedded in substrates, and
also through-the-air. This Chapter analyzes cylindrical glass optical fibers.
Rectangular optical waveguides, used in planar lightwave integrated circuits, are discussed in Chapter 8.

Section 2.2 analyzes the propagation of light in cylindrical optical
fibers. Ray optics, also referred to as geometrical optics, is used to initially
describe propagation in and around the core of multimode fibers. Rays are
an acceptable substitute for solving the complex equations required to
model multimode fiber propagation. Single-mode fiber analysis can use
some of the results of geometrical optics, but requires the use of EM wave
theory in addition. The *step index fiber* is used as an initial model because
it provides a simple platform for analysis. In a step index fiber the index of
refraction change at the core/cladding interface is abrupt. Section 2.2 also
develops the concepts of polarization, total internal reflection, reflection
and reflectance, numerical aperture, and wave velocities using the step-
index fiber model.

Section 2.3 builds on the velocity concepts to explain core propagating
modes and cutoff frequencies. The two basic types of fibers used for telecommunications, *single-mode (SM)* and *graded index (GMM) multimode,* are
described. Different SM core indices of refraction, beyond the step-index, are
considered. Single-mode fibers do not have the problem of modal dispersion.

As a result, their core/cladding interfaces can be optimized to control total dispersion.

Finally, Section 2.4 discusses the manufacture and applications of fibers and cables. The technology has become relatively mature. Current optical fiber production is highly automated and precisely controlled. Costs per fiber-meter are continually declining, particularly for single-mode fiber. A customer can obtain fiber in many different packages and with a range of characteristics intended to satisfy a wide variety of applications.

2.2   Rays & Waves

Consider a point source of light distant by many kilometers. In radio terms this would be considered an isotropic radiator because the EM wave would spread out uniformly in a sphere. The maximum value of the wave can be thought of as a crest traveling outward. As long as the medium is uniform in its characteristics and unchanging, the velocity will be constant. A receiving aperture will see a small portion of the sphere's wave front, which will basically appear as a flat plane. This is referred to as a *plane wave* front, and is pictured in Figure 2-1. The wavelength, λ, also has to be much smaller than the viewing aperture for the assumption of a flat or plane wave front to be valid. For example, light impinging on a very small pinhole would not satisfy this requirement. Diffraction, or bending, of the light at the pinhole's edges will generate an interference pattern beyond the aperture.

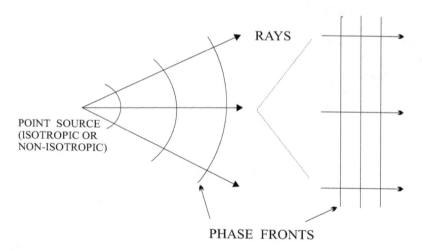

Figure 2-1 *Plane waves.*

A wave front is defined as the locus of all points in the wave having the same phase. As a result, it is also called a *phase front*. If the distance is great, the radiator can be non-isotropic (non-uniform in all directions) and the wave at the small aperture will still be a plane wave.

The apertures or openings in fibers are large relative to the wavelengths of the optical EM waves. As a result, rays or straight lines can be drawn perpendicular to the phase front to describe the propagation of the wave and its energy flow. This allows large-scale optical effects such as reflection and refraction to be analyzed by a relatively simple geometrical process called *ray tracing*. Ray optics is not relied upon to explain (small core) single-mode fiber performance, but the use of rays gives intuitive insights into the propagation in all fibers.

Figure 2-2 uses ray optics to describe what happens at the boundary or interface between two media that have different indices of refraction. Notice the bending, or refraction that takes place in the second material. With $\eta_2 < \eta_1$, the refracted ray is bent away from the normal to the boundary surface. With $\eta_2 > \eta_1$, the refracted ray is bent toward the normal (as pictured in Fig. 2-2). The governing angles are always defined relative to the normal to the boundary, thus ϕ_1 is always labeled the *angle of incidence*. In all optical fibers $\eta_1 > \eta_2$ and the refracted ray is bent away from the normal to the interface not towards the normal. The *angle of reflection* is always equal to the angle of incidence, and these two rays will lie in the

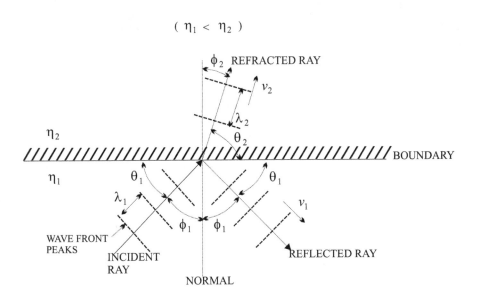

Figure 2-2 *Reflection & refraction.*

same plane. The polarization, or orientation of the E-field vector, relative to this plane is important in determining what happens at the boundary, and also how the wave propagates (or does not propagate) in the fiber.

2.2.1 *Polarization*

An EM wave has two components, an electric and magnetic field. In a vector analysis of the wave, the electric field is labeled $\vec{E}$, the magnetic field $\vec{H}$, and they are orthogonal or 90° to each other. Both are varying with time and distance, as pictured in Figure 2-3. The wave is propagating in the Z direction. The vector is usually associated with the maximum or peak value. The vector orientation of the two fields in an EM wave is always shown transverse to the direction of propagation. The direction of the E field vector is used to describe the wave's *polarization*. If the polarization is fixed in time and space, the wave is linearly polarized. Polarization can also be elliptical or circular, which means that the E field direction is changing with time and/or distance. The energy or power carried by the wave is proportional to the square of the electric field and is often referred to as the intensity of the optical wave. Since coherent generation and detection of optical waves has not yet reached practical usage, the wave's intensity will be used throughout this text. Power is the time rate at which energy is delivered.

In a metallic waveguide, when a wave strikes a metallic interface, the reflected wave may have less amplitude and may encounter a phase change. A standing wave pattern can result, formed by the interaction between the incident and reflected waves. In an optical fiber dielectric

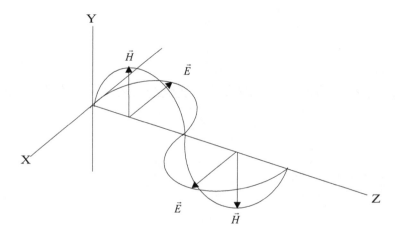

Figure 2-3 *Electric / magnetic fields.*

waveguide, no surface currents at the boundary are possible so a different reflection pattern is generated. Regardless of the type of interface, reflected waves will generally encounter energy loss and a phase shift. The incident and reflected waves produce patterns of propagating waves that depend on the source's radiation pattern, and the input alignment and properties of the fiber (particularly core size relative to a wavelength). These propagating wave patterns are called *modes*.

2.2.2   *Step Index Fiber*

The ratio between the energy transmitted and the energy reflected depends on the incident wave polarization and the differences in the index of refraction of the two materials. Also, since the refractive indices are different, from equations 1.1 and 1.3 the wavelengths and velocities in the two materials will be different. The incident and refracted angles at the interface are related by Snell's law:

$$\eta_1 \sin \phi_1 = \eta_2 \sin \phi_2, \text{ or equivalently,} \qquad\qquad 2.1$$

$$\eta_1 \cos \Theta_1 = \eta_2 \cos \Theta_2$$

Snell's law helps explain the phenomenon of total internal reflection that is critical to optical fiber propagation. Picture the interface as that between the core and cladding of a cylindrical optical fiber, where η_1 has a slightly higher index of refraction than η_2. A fiber of this type is shown in Figure 2-4. In a Silica fiber, the core's index of refraction will generally be only a small fraction greater than the cladding. If the core's index is assumed to be uniform radially and longitudinally, the fiber is referred to as step-index. A step-index fiber has an abrupt change in index at the core/cladding boundary. The values in Figure 2-4 for the core/cladding indices are typical of a multimode, step-index fiber but these values depend on the fiber design and application. The indices of refraction for Silica fibers vary around a value of 1.5. Polymer coatings are often used to protect the cladding. Their indices of refraction are in the same range.

Because η_1 and η_2 are very close numerically, a useful mathematical relationship can be derived based on the fractional change in the indices, Δ.

$$\Delta = \frac{(\eta_1^2 - \eta_2^2)}{2\eta_1^2} \approx \frac{(\eta_1 - \eta_2)}{\eta_1}, \quad \text{from which} \qquad\qquad 2.2$$

$$\eta_2 \approx \eta_1 (1 - \Delta)$$

The cladding in silica fibers is purposefully thick so that the fields guided in the core are close to zero at the cladding/coating (or jacket or buffer) interface. The protective jacket is often color coded to allow easy identification. Light leaking from the core through the cladding will not

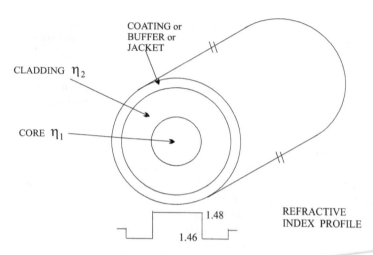

Figure 2-4 *Fiber structure.*

penetrate through the buffering layers unless it has very high energy or the fiber has a sharp bend. The fiber surface is also protected so that flaws will not lead to poor transmission or breakage. In spite of the external jacket protection, fibers must be handled with care at all times. Damage may not be immediately apparent but serious problems can develop over time: think of the slowly developing crack line in a windshield after a pebble's impact.

2.2.3 Total Internal Reflection

In Figure 2-2, the η_1 index is lower than the η_2 index. In optical fiber transmission the situation is reversed. Figure 2-5 shows this combination and how, as the angle of incidence increases, the fiber will allow propagation. Figures 2-5B and C show that once the angle of incidence, ϕ_1, is greater than a critical value, ϕ_c, , there will be internal reflection of the light, and energy will not escape. Succeeding reflections down the guide will give total internal reflection and the desired propagation. Since $\phi_2 = 90°$ for this condition, Snell's law gives the following value for ϕ_c:

$$\phi_1 = \text{incident angle} = \phi_c = \sin^{-1}\left(\frac{\eta_2}{\eta_1}\right) \qquad 2.3$$

A ray striking the interface at ϕ_1 less than the critical angle will have a large part of its energy lost in the cladding. Note that a very small Δ means that the critical angle is close to 90°. This is typical of the core-cladding relationship in silica fibers. It allows the use of *weak guiding*

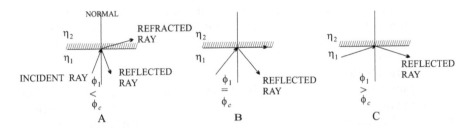

Figure 2-5 *Total internal reflection.*

approximations when propagation is analyzed in detail. In an all-plastic fiber, which will have a large diameter core relative to the cladding, Δ is much greater and the critical angle much smaller. Both types have important applications, which will be discussed in a later section.

Just as in metallic waveguides at the air/copper interface, there will be phase shifts at the optical fiber's core/cladding interface. The amount of shift depends primarily on the angle of incidence, ϕ_1. These shifts contribute to the modal structure of the energy propagating in the guide, but do not impact the reflected energy. In understanding the energy relationships, two parameters are used to quantify a reflection: the reflection coefficient ρ, which is the ratio of the reflected-to-incident electric field maximum values; and the reflectance, R, which is the ratio of the reflected beam intensity (energy) to the incident beam intensity (energy). The reflectance, R, varies from 0 to 1; the value 1 occurring under conditions of total internal reflection in the fiber. When the incident beam is normal to the boundary ($\phi_1 = 0°$), the reflection coefficient is simply:

$$\rho = \frac{(\eta_1 - \eta_2)}{(\eta_1 + \eta_2)} \qquad 2.4$$

Since the intensity is proportional to the square of the electric field, the reflectance becomes:

$$R = \frac{(\eta_1 - \eta_2)^2}{(\eta_1 + \eta_2)^2} \qquad 2.5$$

For an air-to-glass interface, using $\eta_2 = 1.5$ for the glass index of refraction:

$$R = \frac{(1 - 1.5)^2}{(1 + 1.5)^2} = 0.04 = 4\%$$

The transmission loss in decibels is:

$$\text{Loss in dB} = -10\log_{10} 0.96 = 0.18$$

Note that if a glass-air interface were considered, the loss would be the same. This value represents a lower (reflectance based) loss limit on all air and glass couplings and will be used in later sections. When an incident ray is at an angle, the reflection coefficient and, hence, the reflectance is dependent on the incident ray's polarization and the angle of incidence.

For purposes of reference in the following analysis, a plane of incidence is defined as the plane formed from two lines: the incident ray's direction of travel and a normal to the reflecting surface. A perpendicularly polarized wave has it's E-field 90° to this plane of incidence. A parallel polarized wave has it's E-field in the plane of incidence. Any linearly polarized wave of arbitrary polarization can be broken into perpendicular and parallel components. Figure 2-6A illustrates these polarizations for $\eta_2 > \eta_1$, which would be the case for a wave incident from air to glass, for example. Parallel polarization is labeled "P" and perpendicular polarization is "S" in keeping with an existing standard approach. From Fresnel's laws of reflection, the reflectance is found from:

$$R(Parallel\ polarization) = |\rho_P|^2 = \left| \frac{-\eta_2^2 \cos\phi_1 + \eta_1 \sqrt{(\eta_2^2 - \eta_1^2 \sin^2 \phi_1)}}{+\eta_2^2 \cos\phi_1 + \eta_1 \sqrt{(\eta_2^2 - \eta_1^2 \sin^2 \phi_1)}} \right|^2$$

2.6

$$R(Perpendicular\ polarization) = |\rho_S|^2 = \left| \frac{+\eta_1 \cos\phi_1 - \sqrt{(\eta_2^2 - \eta_1^2 \sin^2 \phi_1)}}{+\eta_1 \cos\phi_1 + \sqrt{(\eta_2^2 - \eta_1^2 \sin^2 \phi_1)}} \right|^2$$

2.7

Figure 2-6B gives plots of the reflectance for three important boundaries: core/cladding interface with 1.48/1.46 index difference; glass-to-air

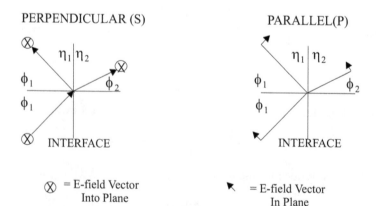

PERPENDICULAR (S)　　　　　　　PARALLEL(P)

Figure 2-6A *Reflections/polarization.*

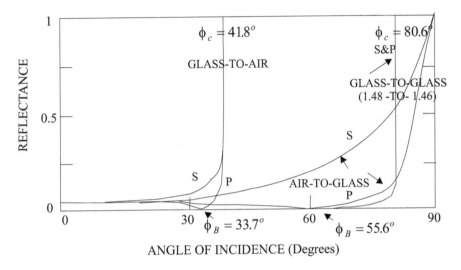

Figure 2-6B *Reflectance.*

with 1.5/1.0 difference; and air-to-glass with 1.0/1.5 difference. The independent parameter is the angle of incidence. Since power is the important parameter in optical fiber intensity modulated systems, not EM field parameters, the phase shift and field magnitudes can be ignored in favor of the reflectance.

Note that the glass-to-air interface has a critical angle of 41.8°, and the air-to-glass interface has no critical angle. With parallel polarization each of them has an angle, called the Brewster angle, at which no reflection takes place. This occurs when the numerator of Equation 2.6 is zero, and is obtained more easily from:

$$\tan \phi_B = \frac{\eta_2}{\eta_1} \qquad 2.8$$

2.2.4   *Numerical Aperture*

One of the most important parameters of a fiber or, for that matter, any optical device that accepts and uses light energy, is its numerical aperture. The *numerical aperture* gives the proportion of impinging light that can be accepted and used. In Figure 2–7 a step index fiber is shown accepting light from a source. Note that some rays from the source will not be guided because they enter the fiber at too great an incident angle. Their incident angle is refracted in the core and strikes the core/cladding boundary at less than the fiber's critical angle. The interface between the outside air and the fiber's glass core and cladding does not have a critical angle because $\eta_1 < \eta_2$.

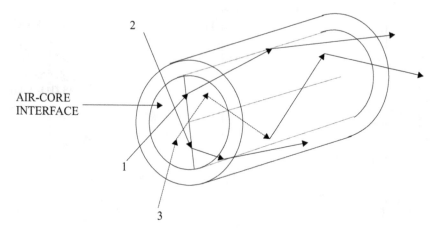

AIR-CORE
INTERFACE

Figure 2-7A *Rays in multimode fibers.*

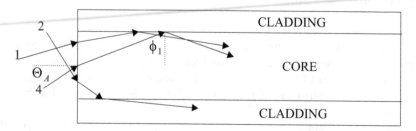

Figure 2-7B *Lateral profile.*

There are two basic types of rays that will propagate, meridian and skew. *Meridional rays* are confined to the meridian (longitudinal) planes of the fiber. The core or fiber axis is contained in each meridian plane. Meridian rays can be guided (Figure 2.7A, #1) or refracted out of the core (Figure 2.7A, #2). *Skew rays* (Figure 2.7A, #3) are not confined to a single plane and follow a helical path down the guide. Skew rays probably make up the greater proportion of the total number of rays, but the meridian core rays are responsible for the longer distance transmission in the fiber. Skew ray #3 will contribute to what is labeled a leaky mode whose energy rapidly attenuates with distance. Refracted ray #2, which is a meridian ray, will contribute to cladding modes. These modes can travel longer distances than leaky modes but will still be attenuated relative to the core modes.

Considering ray #4 from the air outside the fiber, and examining the lateral profile of rays 1,2 and 4 using Figure 2-7B will yield an equation for the fiber's numerical aperture. Ray #4 enters the fiber at incidence angle θ_A, which, when refracted, is assumed to be incident on the core/cladding boundary at the critical angle ϕ_c. Using trigonometry:

since $$\sin \phi_c = \frac{\eta_2}{\eta_1}, \text{ from Equation 2.3}$$

and $$\sin^2 \phi_c + \cos^2 \phi_c = 1$$

thus $$\cos^2 \phi_c = \frac{(\eta_1^2 - \eta_2^2)}{\eta_1^2} \qquad 2.9$$

and from Snell's law, $$\eta_0 \sin \Theta_A = \eta_1 \sin \Theta_1$$

where $$\Theta_A = \text{Acceptance Angle}$$

since $$\eta_0 = 1, \text{ for air outside the core,}$$

and $$\sin\Theta_1 = \cos\phi_1 = \cos\phi_c,$$

$$\sin\Theta_A = \eta_1\cos\Theta_c = \eta_1 \left(\frac{\sqrt{(\eta_1^2 - \eta_2^2)}}{\sqrt{\eta_1^2}} \right) \qquad 2.10$$

$$= \sqrt{(\eta_1^2 - \eta_2^2)} = \text{numerical aperture (NA)}$$

also $$NA = \sqrt{(\eta_1^2 - \eta_2^2)} = \sqrt{(\eta_1 - \eta_2)(\eta_1 + \eta_2)}$$

$$\approx \eta_1 \sqrt{2\Delta} \qquad 2.11$$

The larger the NA, the greater the fiber's ability to collect and transmit light. Multimode fibers will have significantly larger NAs than single-mode fibers because of their larger cores. This helps significantly with optical source and connector coupling. However, they will also have the modal dispersion that is absent from single-mode fibers. For the step index fiber the acceptance angle, and hence the NA, is constant across the face of the core. Later it will be shown that other fiber types will have variable NAs across the core's face because their cores have shaped refractive indices.

2.2.5 *Velocities*

As pictured in Figure 2-2, the EM wave has maximums at the peak values and, as a result, minimum values between the peaks. At the boundary between η_1 and η_2 the wavefront peak can be thought of as an ocean wave hitting a beach. In Figure 2-8 a second wavefront is moving along the (beach)

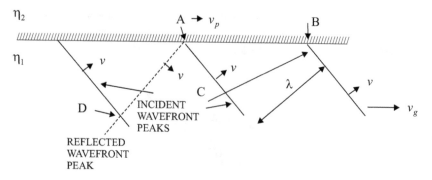

Figure 2-8 *Waveguide velocities.*

boundary at point A with a phase velocity, v_p. The next wave peak hits the boundary at point B, which is also moving at the same phase velocity. The wavelength, λ_p, between them is longer than the free space wavelength, λ, which means that the phase velocity is higher than the inherent wave velocity, v. The velocity of a plane wavefront in an infinite medium is equal to both the wave velocity and phase velocity. Recall that the basic wave velocity is dictated by the frequency and the index of refraction according to $\lambda = v/f = c/(v_1 \times f)$. If the wave makes a very small angle with the boundary (very large incident angle), the resultant phase velocity will be very large. If the wave's incident angle is zero, the wave is normal to the surface.

In Figure 2-8, as the point A on the boundary moves, a reflected wavefront is launched with the same wave velocity, v, as the incident wave. What happens to the relative phasing between the incident and reflected waves at this interface depends on the state of polarization of the incident wave. Some of the incident wave's energy might also be dissipated at the boundary because of the transmitted (refracted) wave. The two peaks, incident and reflected, meet at point D, which is constantly moving. If the boundary provides total reflection with no phase shift, the vector sum of the two velocities would be directly down the guide. A pattern or structure of propagating EM maximums and minimums will be formed in the guide regardless of how the reflected wave is affected. These are called modes, and the modal structure depends on the size, structure, and material in the guide, as well as the frequency of the incident wave. The modes propagate at a velocity called the group velocity, v_g, which is always less than either the wave and phase velocities. The group velocity is the rate at which energy is propagated in the confining waveguide structure. In free space, a plane wave has a group velocity equal to the speed of light, c. In all waveguides v_g is lower than the speed of light because the boundary reflections generate the slower moving modal structures. The group, phase, and free space velocities are related by:

$$\nu_g \nu_p = \frac{c^2}{h_1^2} = \nu^2 \qquad\qquad 2.12$$

If two different propagating frequencies in a guide are considered, one lower in frequency than the other, EM analysis (not included here) shows that the lower will have smaller incident and reflection angles. In both cases the angles of incidence and reflection are greater than the critical angle or core propagation would be impossible. From the Equation 2.11 relationship, with a large incident angle the phase velocity will be high and the group velocity low. Since the transmitted information moves at group velocity and is always made up of a combination of frequencies, the constituent frequencies within a signal's spectrum will arrive with different delays. This result, referred to as *waveguide dispersion*, is present in all guiding structures to a greater or lesser degree.

When modal propagating patterns exist in the fiber, the higher order mode rays will strike the core/cladding interface at greater incident angles. The modes with greater angles will arrive later, leading to *modal dispersion*. The problems caused by waveguide and modal dispersion are discussed more thoroughly in Chapter 3.

2.3   Modes

There are other points of maximum and minimum strength than just point D in Figure 2-8. This leads to the conclusion, without detailed analysis, that within the guide there is a moving or propagating structure of EM fields or modes. The cylindrical optical fiber could be modeled and solved for these propagating modes using Maxwell's equations but the math is too complex for this treatment. Ray tracing and certain results from EM theory will provide a practical understanding of the propagation phenomenon and give a sufficiently clear picture of the differences between multimode and single-mode fibers.

2.3.1   *Step-Index Modes*

Certain EM wave boundary conditions must be met at the interface between two different waveguide media for propagation to be sustained. These media can be an air-filled, rectangular copper guide for RF, for example, or the core and cladding of an optical fiber (Figure 2-4). The main difference between them is that the fiber has a dielectric interface while the copper waveguide has a metallic interface. The metallic waveguide essentially has no fields in the walls of the guide, while the fiber has fields extending into the cladding. For a propagating wave, only certain combinations of EM field structures, called modes, will meet the conditions posed by a given waveguide. The fiber will have a more limited mode structure in the core than the metallic waveguide because modes caused by currents in the guide walls are not present. On the other hand, the fiber will allow some modes to propagate in the cladding, generally only for short distances. Core modes have most of their energy traveling longitudinally in the fiber.

The low-order modes for fibers have transverse (to the propagation direction) electric and magnetic fields. These are labeled $TE_{x,y}$ and $TM_{x,y}$ (transverse electric and transverse magnetic), where the x and y refer to the number of electric and magnetic field variations transverse across the guide. In addition hybrid modes will propagate, which are combinations of TE and TM modes and have vector components along the axis of the core. These are labeled $HE_{x,y}$ and $EH_{x,y}$ modes. Some modes are linearly polarized and labeled LP modes. Higher order modes have slower group velocities, which causes modal dispersion at the receiver.

The E-field structures for the fundamental and the first set of (low order) modes are illustrated in Figure 2–9 to indicate what is happening in a step-index cylindrical guide. A rectangular optical guide, often found in optical ICs, will have a somewhat different modal structure. Because of symmetry in the circular fiber, these patterns actually represent two modal structures each, one orthogonally polarized relative to the other. This is why the modes are also referred to as $LP_{x,y}$.

ELECTRIC FIELD VECTOR

Lowest Order (Fundamental) Mode

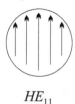

HE_{11}

First Set of Higher Order Modes - LP_{11}

TE_{01} TM_{01} HE_{21}

Figure 2-9 *Step waveguide lowest order modes.*

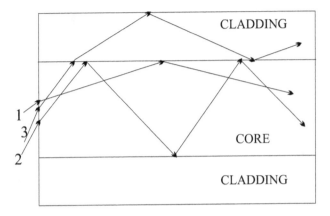

Figure 2-10 *Step-index modes.*

Ray tracing can be used to help explain the behavior of high-order modes, but is not applicable for completely describing the fundamental, HE_{11}, mode. The fundamental and higher order modes propagating in the core will have energy traveling in both the core and cladding. Figure 2-10 illustrates with ray geometry some of the path differences between high-order core modes, low-order core modes, and cladding modes in a cylindrical, multimode, step-index fiber. Ray 1 represents the fundamental mode. Its effective index of refraction is close to that of the core index, η_1. Ray 2 represents a higher order propagation mode. Its effective index is close to that of the cladding η_2. Ray 3 represents a cladding mode which will die out after a limited distance. The rays show the differences in total distances traveled, and an obvious conclusion is that the information on the modes reaches the receiver at different times. Cladding modes can travel a fairly long distance, but eventually die out relative to the core modes. As discussed in Chapter 3, the modes will mix, which leads to a stable and predictable modal performance if the fiber is long and undisturbed. Skew modes, caused by skew rays, travel in the core and cladding. They die out quickly after the point of generation.

2.3.2 *Cutoff Wavelength*

The modal performance in the core of a cylindrical, step index, optical fiber is determined by a number of parameters: radius a; free-space wavelength λ; core refraction η_1; cladding refraction η_2. If the wavelength is large relative to the radius there will be fewer modes. If the core's index is much larger than the cladding's index there will be more modes. A parameter, called the *normalized frequency V*, combines all of these variables to afford a simplified analysis of fiber designs.

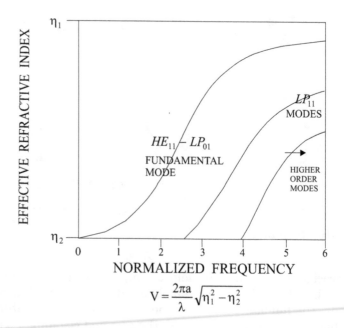

Figure 2-11 *Step index modes.*

A mode chart, Figure 2-11, plots the relationships between some of these modes or *mode groups* as a function of V. A mode group is made up of modes that travel at the same group velocity. A properly designed multimode fiber will have a large normalized frequency and sustain hundreds of individual modes in the core. The vertical axis gives a parameter called the mode's effective refractive index, which will lie between that of the core and cladding. Given a more complex analysis, this parameter would be replaced by an effective propagation coefficient, but the resulting comparative performances would not change. Since group velocity is inversely related to the refractive index, the chart basically shows the relative speeds of the various modes. The LP_{11} mode group will not propagate if V is less than 2.403. This is the *cutoff wavelength,* and determines the region of single-mode propagation.

A large difference in modal speeds will yield significant modal dispersion at the detector. The step-index modal performance can be improved upon by shaping the core's refractive index in the form of a parabola. This type of multimode fiber is referred to as *graded index, or GRIN* fiber (discussed later in Chapter 3). From Equation 2.3, a fiber's critical angle will decrease with an increasing core index relative to the cladding, thus allowing more rays or modes to propagate. Modes are cutoff when their rays travel at the critical angle. Modes far from cutoff will have incident angles close to 90°, thus traveling almost directly down the guide but at a slower velocity. A high-order mode operating far from cutoff will have an effective index close to the core's index and thus will be slowed ($vel = c/\eta$). For any given mode, the effective refractive index varies with wavelength (V changes as λ changes), which gives rise to a phenomenon called waveguide dispersion (Chapter 3).

On the horizontal axis in Figure 2-11 the four parameters critical to the fiber's design are combined into the normalized frequency variable V. The normalized frequency is defined as:

$$V = \left(\frac{2\pi a}{\lambda}\right)(\eta_1^2 - \eta_2^2)^{1/2} = \frac{2\pi a \eta_1}{\lambda} \sqrt{2\Delta} = \text{normalized frequency} \qquad 2.13$$

where a = radius of curvature of core

If V is small, the number of core propagating modes can be counted on the mode chart. If $V \geq 10$, the number of core propagating modes in a step-index fiber can be estimated by:

$$N = \frac{V^2}{2} = \text{number of modes} \qquad 2.14$$

When the wavelength of a propagating higher order mode or mode group is large relative to the core diameter the normalized frequency is reduced and the mode or mode group may be cutoff; i.e., not able to propagate in the core. Note that the fundamental $LP_{0,1}$ mode will always propagate.

As the normalized frequency for a particular mode group approaches cutoff from the high side, more of the power is transferred to the cladding, where it is lost in a relatively short longitudinal distance. Energy in the cladding is susceptible to external effects, such as pressure or bends. At cutoff the mode will exist only in the cladding, and it ceases to propagate in the core. For large values of V, the fraction of total power found in the cladding can be estimated from:

$$\frac{P_{cladding}}{P_{total}} = \frac{4}{3\sqrt{N}} \qquad 2.15$$

2.3.3 *Single Mode Fiber*

There are two basic types of optical fibers used for telecommunications: single-mode (SM), and multimode (MM). They can be identified mainly by the size of their cores. Table 2-1 lists some of the characteristics of different fibers. Note that some of these are made of plastic, not glass. Plastic fibers are best suited for optical sources in the visible light region. Plastic core

Table 2-1 *Typical Fiber Characteristics*

FIBER TYPE	CORE TYPE	η_1	η_2	CORE DIAM. in μm	NA	ACCEPTANCE ANGLE – in degrees	3 dB BW-1km in Mhz
Silica	Step MM	1.48	1.46	50	0.24	13.9	7.7
Silica	Graded MM	1.48	1.46	50	0.24 variable	13.9 variable	44,000
Silica	Step SM	1.465	1.461	8	0.1	5.7	500,000 @ 1500nm 0.1 nm linewidth
Plastic Coated Silica	Step MM	1.46	1.4	200	0.41	24.2	2.5
Plastic	Step MM	1.49	1.42	1000	0.46	28	2.1

fibers are usually step-index. Glass or silica fibers will use either step or shaped index profiles. The characteristics and applications of most of these fibers will be discussed in detail later.

If the core of the cylindrical fiber is sufficiently small relative to a propagating wavelength, the fiber will support only the fundamental mode. Relatively large core fibers, by contrast, are designed to allow many modes to propagate. As a rule, fibers are designed to support either the fundamental mode or a fairly large number of modes. Silica optical fibers are designed to operate in three wavelength regions between 0.8 microns and 1.6 microns. As a result, SM cores will have diameters of 5–10 microns and MM cores will be greater than 50 microns. The very large core of the plastic fiber is needed so that more light can be accepted, thus offsetting the high attenuations in plastic. Typical attenuations are not mentioned in this chart and will be discussed later. The SM fiber gives the best bandwidth (and attenuation) performance but the small NA and acceptance angle lead to greater coupling problems.

2.3.4   *Step-Index Single Mode Radius*

From Equation 2.13, if a combination of the four basic fiber parameters results in $V \leq 2.405$, the step-index fiber will be single-mode. In practice a SM fiber will be designed with $V \approx 2.0$ so that there will be a minimum of multimoding at fiber discontinuities. For the step-index, SM fiber, the fundamental mode intensity is circularly symmetric and given by a Gaussian shape:

$$\text{Intensity} = \text{Power Distribution} = I = I_{\text{max}}e^{-2r^2/w^2}, \text{ where} \qquad 2.16$$

$$r = \text{distance from center and}$$

$$w = \text{spot size} = 13\frac{1}{2}\% \text{ point on Gaussian curve}$$

and $\qquad I_{max} = \text{center intensity}$

The *spot size* definition is convenient because it is the circle on the intensity distribution where the energy falls to $1/e^2$ of the maximum value. The spot size measures the mode field and helps determine how much light energy can be coupled out of an SM fiber into other optical devices, like detectors and connectors. At $V = 2$ about 75% of the power is in the core, while at $V = 1$ only 30% is in the core. As a result, with too low a normalized frequency a SM fiber runs the risk of excess losses at connectors, splices, etc.

2.3.5 Shaped SM Fiber Cores

Most SM fibers currently in production use core/cladding interfaces more complex than a simple step-index. Figure 2-12 shows approximations of the cross-sections of some of these refractive index profiles. Recall that SM fibers have only waveguide and material dispersion. Dispersion effects can be partially mitigated by controlling the core profile. This controls the fundamental, LP_{01}, mode waveguide dispersion which in turn can be used to compensate for material dispersion. Figure 2-12A illustrates a depressed cladding, step-index SM fiber that allows the loss due to core dopants to be lowered. Note that the cladding is wide which is intended to present the core fundamental mode with a fixed value of cladding that extends essentially to infinity. The outer index is that of the glass used to begin fiber fabrication. Often this is a tube of pure Silica. Figure 2-12B gives the cross-section profile of a dispersion-shifted fiber. This fiber has a much narrower depressed cladding and a shaped core. The value of shifting the total dispersion curve is discussed in Chapters 3,7, and 8. Figure 2-12C shows a dispersion-flattened profile.

The normalized frequency at higher-order mode cutoff for the shaped SM fibers is different than that for the step-index. Also, the mode field diameters differ relative to that of a step-index fiber. The dispersion-flattened mode field diameter is within about 5%, but the dispersion-shifted mode field varies by a much greater amount.

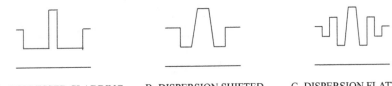

A- DEPRESSED CLADDING   B- DISPERSION SHIFTED   C- DISPERSION FLATTENED
STEP INDEX

Figure 2-12 *Single-mode fiber index profiles.*

2.2.6   *Graded Index Multimode Fiber*

The core index profile for a graded index (GMM) multimode fiber is not a step. It gradually changes from the core central axis maximum value, η_1, to the cladding value, η_2. The optimum shape has been found to be approximately parabolic. A mathematical model for a parabolic core profile, assuming circular symmetry is given by:

$$\eta(r) = \eta_1 \sqrt{1 - 2 \left(\frac{r}{a}\right)^2 \Delta} \quad for \; r < a \qquad\qquad 2.17$$

$$= \eta_2 \qquad\qquad\qquad for \; r \geq a$$

where a = radius of core, r = radial distance from axis, and $\eta_1 = \eta_2$.

Note that at the central axis, $r = 0$ and $\eta(0) = \eta_1$ while at the cladding $r = a$

and

$$\eta(a) = \eta_1 \sqrt{(1 - 2\Delta)} \approx \eta_1 (1 - \Delta) = \eta_2.$$

Guiding of the core modes is not done by total internal reflection at a single boundary as with the step index fiber. Waves in the core follow a sinusoidal path depicted in Figure 2–13. In the model the parabolic index shape is approximated by a number of small changes from layer to layer as the core index decreases to that of the cladding. In fact, GMM fibers are indeed manufactured with just that technique: many layers of small step changes are deposited to build the parabolic core shape. Using ray optics, Figure 2-13 sketches a portion of the paths for a high and low order mode.

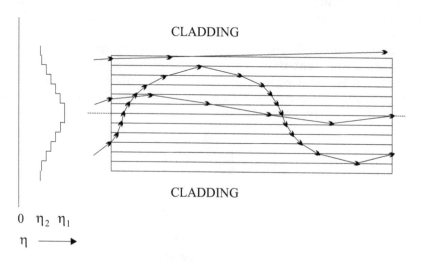

Figure 2-13   *Graded index modes / rays.*

A ray from a source is shown entering the core near the cladding at a very small angle relative to the air/core interface. Using the earlier angle definitions, this would be an incident angle close to 90°. Notice that it still escapes into the cladding. The GRIN fiber has a variable NA as a result of the parabolic refractive index shape. At the center the NA is found using Equations 2.10 and 2.11. At $r = a$, the radius of the fiber core, the NA is equal to 0. An equation for this variation of NA as a function of the radius is:

$$NA = \eta_1 \sqrt{2\Delta} \sqrt{\left(1 - \left(\frac{r}{a}\right)^2\right)} \qquad 2.18$$

when $r = 0$, $NA = \eta_1 \sqrt{2\Delta}$, the same as a step-index.

As with the step index fiber, the high order modes will travel. However, the travel times of the modes will be closer together because, with the GRIN fiber, the higher order modes will be traveling in a region of lower refractive index. Thus the longer distance and travel time of higher order modes is partially compensated by greater velocity ($v = c/\eta$). This leads to lower modal distortion for the GRIN fiber relative to the step-index fiber because the velocities of the propagating modes have reduced spread.

The normalized frequency parameter, V, for a GRIN fiber is obtained using the same equation as the step-index fiber (Equation 2.13). The number of modes in a multimode GRIN fiber is reduced, and approximated by:

$$N = \frac{V^2}{4} \text{ for } V > 10 \qquad 2.19$$

This is half the number found in a step-index fiber with the same normalized frequency.

2.4   Optical Fiber/Cable Manufacture

The primary materials used to make window glass are silica (SiO_2), calcium oxide (CaO), and sodium oxide (Na_2O). Even though window glass appears transparent, its attenuation of about 1000 dB/km makes it unsuitable for long distance optical transmission. The starting ingredient in glass optical fibers is purified Silica. In pure Silica the contaminant concentrations that would absorb energy in the range of 600 nm to 1600 nm have been reduced to levels of about 1 part per billion. The pure silica used in fibers has atoms arranged randomly, not in the patterns seen in crystals. Quartz is crystallized silica, and is not useful for optical fibers.

The dependence of the index of refraction of silica on wavelength is plotted in Figures 3.3 and 3.4. Also shown is the dependence of the refractive index of silica when dopants are added. The dopants are added when optical fibers are made to obtain the desired core/cladding profiles. This determines the fiber's attenuation and dispersion performance as a function of wavelength. Since the fiber's core has to have a higher index than the cladding, the primary core dopant for most Silica fibers is Germanium (Ge). A dopant often used to make the cladding index lower is Flourine (Fl)

One approach to making fibers uses a traditional glass production method, the double crucible. Rods of silica with the desired doping are fed into two concentric, tapered chambers: the inner for the core and the outer for the cladding. The glass in the rods is melted and the two types merge to form a core and cladding. The fiber is drawn at the crucible's neck in a continuous process. The costs are low, but the method cannot produce the complex core/cladding cross-sections required for modern SM systems.

The methods favored currently use deposited vapors at high temperatures to make preforms. A basic three-step process is used to turn pure Silica into cabled fibers by the preform method:

- A cylindrical *preform* of a few centimeters in width and about a meter in length is manufactured. The preform is a large version of the desired fiber, with the core and cladding doped by using different gases. The glass industry also refers to preforms as *boules*.
- A fiber is drawn from the preform.
- Fibers are individually buffered and assembled in cable structures.

Preforms are made using a number of different vapor-phase oxidation and deposition processes. Oxygen (O_2) and vapors of silicon tetrachloride ($SiCl_4$) and/or germanium tetrachloride ($GeCl_4$) are injected into a heated environment under precise computer control. The reaction leaves controlled, deposited layers of soot that are either germania (GeO_2) + silica or pure silica. A core doped with germanium, plus a pure silica cladding, can give any desired fractional refractive index difference, Δ. The soot is collected either internally or externally on starter silica tubes or rods. The leading (proprietary) methods used to make preforms can be described as: outside vapor deposition (OVD), axial vapor deposition (AVD), modified chemical vapor deposition (MCVD), and plasma chemical vapor deposition (PCVD).

Other dopants than Ge are used depending on the type of fiber being manufactured, and the desired core/cladding cross-section. Flourine (Fl) and boron (B) doping will lower the refractive index. A *depressed cladding* fiber uses a flourine doped cladding and a core with reduced germanium doping. The attenuation is slightly lower compared to a fiber with a pure silica cladding.

The second basic step is to draw a fiber from the preform. The preform is an enlarged version of the fiber. Figure 2-14 sketches a *fiber drawing tower*. The end of the preform is melted and a fiber in kilometer lengths is drawn from it. A thin plastic coating is applied during the draw. The whole procedure is computer controlled. The core profile was set during the pre-

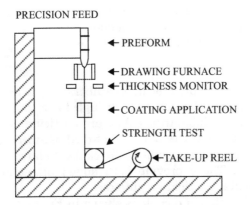

Figure 2-14 *Fiber drawing tower.*

form stage, but the fiber's external quality is controlled during this process. Drawing speeds are in the range of a few meters/sec. Tension testing to a limit of 100,000 psi is performed at the time the fiber is drawn. If the fiber has the common 125 μm outer diameter, without coating, the force required translates to only 1 kg (2 lb). At this stage the fiber is still fragile. Further protections are added during the cabling procedure that follows.

The first step in the cabling procedure is to add another coat of plastic protection, called a buffer. There may be multiple layers of buffering applied, depending on the intended application. The outer buffer is color coated so the individual fiber can be easily identified. Buffer coatings can raise the outer fiber diameter to 250–900 μm. For example, jumper fibers used indoors require thick buffers because they do not have the added protection of additional cabling. Multimode fiber jumpers are generally colored orange and single-mode, yellow. Inside the cable a fiber will be either *tightly buffered* or *loosely buffered*. The latter requires protective tubing because the fiber has to be protected from external disturbances or damage. Anticipated environmental factors dictate the type of buffering chosen.

Two main areas of consideration determine the choice of cable: intended application, and the environment. There are a large number of applications for fiber cables:

- Instrumentation, including special industrial data/telephone
- Inside, office-type telecommunications connections
- Multi-fiber cables in buildings for ducts, overhead plenums, or risers
- Short, exterior connections between buildings (campus, LAN)
- Power line (hybrid-electricity+telecomm.) transmission
- Residence distribution of broadband services

- Short-haul, metro area data/telephone
- Long-haul, high capacity data/telephone
- Ultra-long-haul, submarine data/telephone
- Sensors

Some of these use only single or double fibers (e.g. sensors, instrumentation, office). Others require cables with high fiber counts (e.g. residence, short-haul, long-haul). Submarine cables are particularly interesting. They generally have very few fibers but a great deal of external protection. The unique environment requires submarine cables to have multiple layers of steel/copper/plastic. These provide both added protection and electrical power for intermediate optical amplifiers.

A generalized sketch of a cable is shown in Figure 2–15. The diameter would be about 20 mm. Cables generally range from 10–30 mm in diameter. This cable might possibly be used for campus/metro applications because it has mixed fibers. The total fiber count is 24, separated into 6 groups of 4 fibers each. Three of the groups contain multimode fibers and the other three, single-mode. *Loose-tube* construction is used so that the fibers can move and thus be protected from bends caused by outside forces. The buffering diameters of fibers in loose-tube construction are relatively small. The tubes, as well as the individual fibers, are color-coded. Aramid or Kevlar fibers are used between the tubes and the jacket to add strength and additional protection. A central strength member is used that could be made of metal or a plastic. If the latter, the cable would be all dielectric which would give protection from lightning strokes and ground loop currents. Ground loop currents are caused by potential differences that exist over long sections of (nearby) metallic cable.

In Figure 2-15, the cable interstices and the tubes are filled with a gel to keep out water. Silica fibers are susceptible to water- induced attenuation at certain wavelengths. If a fiber takes on added hydrogen, from water or chemical activity, the attenuation worsens.

This is just one of many environmental affects to consider when choosing a cable structure. They include:

- Moisture incursion
- Animal attacks (e.g. sharks, fisherman, rodents)
- Lightning strikes
- Pressure (e.g. installation pulls, water, traffic, shifting soil)
- Temperature extremes
- Ice load (aerial cables)
- Possible fire exposure
- Harsh chemical exposure
- Need for sharp bending capability

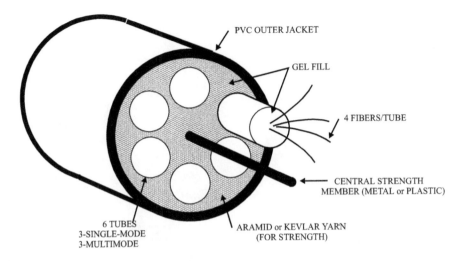

Figure 2-15 *Optical fiber cable.*

Sheathing and internal gels are used to limit water incursion. Reinforced protective jackets are used to guard against critter attack. If lightning or ground loops are a concern, the cable can be all dielectric. Aramid or Kevlar yarns are used in woven fiber form in place of metals to provide strength. Protective jackets and loose tubes are used to protect against pressure and temperature extremes. Cable ducts are also a good protective approach. Extra armoring is used on submarine cables. Messenger wires, to which a cable is attached, are used to protect against some external forces like ice loading. Fibers intended for interior installations in plenums must be coated with a fire retardant material like Teflon. Specially designed jackets are available for harsh environments. If a cable is going to experience sharp bends, the cable diameter should be limited. A rule-of thumb calls for turn radii to be > 20 times the cable diameter if under physical loading, and 10 times if not loaded.

A different type of construction, called a *ribbon cable*, assembles 12 tightly buffered fibers into twelve rows. The big advantage with this type of cable is the potential cost savings during splicing. Since the positions of the fibers are fixed, the splicing operation can be highly automated.

3

Optical Fiber Performance

3.1   Introduction

Popular thinking holds that optical fibers do not have transmission degradations. In reality there are three basic phenomena that govern their performance and must be taken into account when designing optical fiber systems:

- diminished levels of light at the optical detector caused by *attenuation* or loss of signal energy
- time of arrival differences between the different wavelength components of the signal, referred to as *delay distortion* or *dispersion*
- inability of conventional Silica fibers to maintain *polarization*

The first two, attenuation and dispersion, have been a concern since the beginning of optical fiber use. Polarization performance has become a concern as digital signaling speeds on fibers have progressed beyond 2.5 Gb/s. In designing optical fiber systems, end-to-end performance is generally governed by one or the other of the first two degradations, but generally not both at the same time. Polarization degradations become an added source of distortion for high-speed systems.

In this chapter, silica fiber attenuation is discussed first in Section 3.2. The two main types of silica fiber, multimode and single-mode, do not differ greatly in attenuation. Section 3.3 discusses the wavelength dependent dispersion of silica fibers. This results in a time spread at the receiver, $\Delta\tau$, of components within the signal. Multimode and single-mode fibers differ markedly in dispersion performance. This difference determines their two distinct areas of application: local networking, and long haul. Section 3.4 discusses the polarization performance of silica fibers. Finally, Section 3.5 presents information on the performance of non-silica fibers.

The discussion on attenuation and dispersion that follows assumes the fiber transmitting the signal is part of a linear system. For a system, or system component, to be linear the following has to be true: when two independent signals are input simultaneously the output must be the same as the sum of the two outputs that would result if they were input alone. In a linear system the effect on a signal of one component in the transmission path is independent of other components, and also of the presence of any other signal. A fiber will be non-linear under certain conditions, the most obvious being when too much power is input. Later discussions on solitons and optical amplifiers will show the positive advantages of some optical fiber non-linearities.

The time function of a signal transmitted by a linear system can be described in the time domain as $s_o(t)$. The signal has a related (Fourier Transform) spectral function $S_o(\omega)$. Both of these can be distorted in amplitude and phase by a linear system. The system will have a complex transfer function, $H(\omega)$, where:

$$H(\omega) = \frac{S_o(\omega)}{S_i(\omega)} = |H(\omega)|e^{-j\beta(\omega)} = Ke^{-\alpha(\omega)-j\beta(\omega)}, \text{ where} \qquad 3.1$$

$S_i(\omega)$ = input signal,

$\alpha(\omega)$ = frequency dependent attenuation

$\beta(\omega)$ = frequency dependent phase

This linear system assumption is an important first step in analyzing fiber systems. Later chapters in this book will reveal more complex, non-linear behavior in optical fibers. Amplitude distortion is absent when $\alpha(\omega)$ is flat with frequency. Phase distortion is absent when $\beta(\omega)$ is a straight-line function of frequency (wavelength). Because transmission media always have some frequency (wavelength) performance dependencies, the amplitude function will never be flat and the phase function will never be a straight line. The slope of the phase function depends on the velocity of energy propagation. From Chapter 2, in a fiber waveguide this velocity is called the group velocity. In the absence of distortion, the group velocity will be constant with wavelength.

In radio systems, amplitude and phase distortions within the transmitted spectral bandwidth are important in determining total system performance. This is because of the coherent nature of the total signal, which is a carrier plus modulated sidebands. The amplitude and phase relationship of the sidebands to the carrier must be preserved. Current optical fiber systems use only intensity modulation. An intensity-modulated signal carries the baseband signal by modulating the power of the optical source. The transmitted optical spectrum will have energy in a wide range of frequencies (wavelengths), determined by the spectral width of the source. Amplitude and phase distortions within this band do not affect the modulated baseband signal because the spectral wavelengths generated by the carrier are not coherently related to a central frequency. Attenuation of the total power of the optical signal is important, however. At the receiver the total incoming power must be well above any interference and noise. Also, the phase characteristic $\beta(\omega)$ of the fiber will affect the propagation speed of the carrier's wavelengths. A difference in arrival times of the carrier's wavelengths leads to dispersion distortion of the baseband signal.

Recently developed very high-speed systems using near-single frequency lasers require a different view of the modulated spectrum. An intensity modulated single-frequency source will have a spectral width similar to that of a double-sideband AM signal. With a very narrow carrier width, the transmitted optical spectrum will now be determined by the baseband spectral width. In-band amplitude and phase distortions will still not have to be considered because intensity modulation is still used. This advance is analyzed in Chapter 8. At some future date, when optical sources are as coherent as radio oscillators are now and heterodyne reception can be used, the in-band attenuation and phase will become important in determining system performance. In the electrical signal processing portions of a intensity modulated optical system, baseband attenuation and phase distortions still have to be considered.

3.2   Attenuation

Attenuation is found in almost all optical system components. The fiber is the largest contributor, particularly when maximum transmission distance is a primary goal. Other attenuation contributors between optical transmitter output and optical detector input include: optical source and detector fiber coupling losses; optical networking device losses; connector and splice losses; and a loss margin for unknown loss contributors. A system that just meets a predetermined loss design objective that includes all of these contributors is said to be *power* or *loss limited* (Chapter 7). Losses are almost always expressed in decibels since the use of a logarithmic scale simplifies communication link calculations.

When viewed over their useable range of wavelengths, 600–1600 nm, silica fibers have significant attenuation variation with wavelength ($\alpha(\omega)$). The effect of this shape on an intensity modulated signal using a portion of that spectrum is insignificant. As an example, consider the bandwidth required to transmit the signal from a 1550 nm diode laser, intensity modulated by a 2.5 Gb/s pulse stream. The laser is essentially a non-coherent source (Chapter 4) and its resulting average *linewidth* becomes the transmitted linewidth. The instantaneous center wavelength varies greatly throughout this range. Sidebands in the conventional radio sense do not exist, the optical carrier is not sufficiently stable and sidebands are smeared. A signal like this is incoherent in both space and time. Current commercial communication system receivers are looking only for variations in total intensity to demodulate the signal, not the inter-relationship of frequency components (sidebands) within the signal spectrum. These optical power variations will be occurring very rapidly in a 2.5 Gb/s signal, so the detector and receiver have to be designed to have good high frequency response.

The levels of loss generally obtainable in a silica single-mode fiber are shown in the Figure 3-1 attenuation curve. A multimode fiber in the steady state will have slightly higher losses because the higher order (propagating) modes have energy closer to the outer edge where it is more easily lost. The term *steady state* refers to the stable propagating modal structure which is found a few meters distant from the signal launch point.

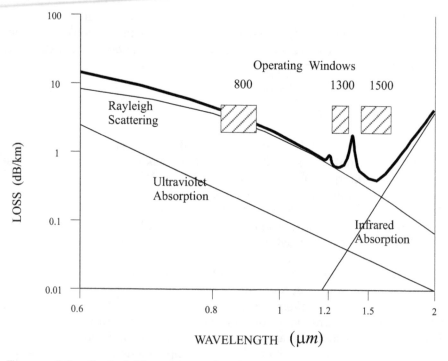

Figure 3-1 *Optical fiber attenuation (silica).*

At short wavelengths ultraviolet absorption is important but is generally overshadowed by *Rayleigh scattering*. At higher wavelengths the minimum attainable attenuations are controlled by *infrared (IR) absorption*. The "V" shaped attenuation curve that results from these two controlling factors, Rayleigh scattering and IR absorption, shows that attenuation is lowest at the relatively longer wavelengths between 1300 and 1550 nm. In Section 3.3 it will be shown that this range fortuitously coincides with the wavelengths where chromatic dispersion is minimized in Silica fibers.

Additional factors, such as impurities and bends, will also contribute to fiber attenuation but they can be minimized with good fiber and cable design, manufacture and installation. Fibers made with different materials, such as plastic, will have higher attenuations. Plastic fibers are attractive in the region of visible light where their loss is minimized. The attenuation is still much higher than that of silica.

Systems using silica fibers have a wide range of sources and detectors available to match the different types and quality of fibers. Through research there is a promise of even lower loss fibers made from materials other than glass or plastic, but the silica based fibers currently in widespread communication use give very satisfactory performance. The supporting hardware and systems that have been developed based on silica fiber use make near term obsolescence highly unlikely. The best values of attenuation for silica glass fibers are 0.15 dB/km for single-mode and 0.24 dB/km for multimode fibers, measured at 1550 nm in the trough between the Rayleigh scattering and IR absorption curves.

Three "windows" of operation for fiber systems evolved during the first two decades of silica fiber use, primarily because of the shape of the attenuation curve. These are the 800, 1300, and 1550 nm windows identified in Figure 3-1. An OH⁻ (water ion) energy absorption peak in silica, centered at 1390 nm, led to the separation of the two longer wavelength windows. Many devices that are used to provide optical output, detection and networking capability are available commercially for these windows. One reason these windows in glass fibers have become standard stems from the chromatic dispersion minimum for pure silica around 1300 nm (Section 3.3). With the gradual improvement in quality of single-mode fibers, it has become possible to use the whole range from 1300 to 1550 nm. In fact, this range now is described using sub-ranges much as radar systems have "bands" in the GHz range (Chapter 8).

The attenuations in Figure 3-1 should be thought of as causing a loss in level to the whole optical signal as it progresses down the fiber. Any in-band attenuation variations will not affect the intensity-modulated signal. Fiber dispersion caused by phase shift variation with wavelength will play a role, however, in amplitude performance by causing rolloff of the fiber's electrical (baseband) response. The larger the dispersion across the signal's bandwidth, the more pronounced this effect. This results in the standard fiber specification given in Mhz-km that describes a fiber's useable bandwidth.

3.2.1 *Material Absorption*

Small amounts of light will be absorbed in an optical fiber because of the chemical composition of glass. The energy is converted to heat. The fundamental mechanism causing this is the excitation of molecular modes of vibration called *resonances*. At short wavelengths, in the ultraviolet region, intrinsic absorption in silica is high due to electronic resonance around 140 nanometers. This loss contribution decreases with increasing λ following $e^{k/\lambda}$. At longer wavelengths intrinsic (IR) absorption is high around 8 μm, also due do molecular vibration, but decreases with decreasing λ following $e^{-k/\lambda}$. These fundamental material absorption characteristics shape the attenuation trough that makes silica fibers attractive. The spectral attenuation plot in Figure 3-1 shows the low wavelength, ultraviolet absorption and the high wavelength, IR absorption. The former is less of a contributor to attenuation because Rayleigh scattering effects are greater at lower wavelengths. Absorption in the IR limits the use of higher wavelengths in all silica fibers.

There are other sources of material-caused attenuation in addition to these intrinsic sources. Very early in the development of optical fiber technology it was realized that metal impurities (e.g., ions of chromium, cobalt, copper and iron) were particularly troublesome. Around 1970 the ability to drastically reduce metal ion concentrations in silica was realized (Figure 1-3). Optical fiber attenuation performance has been improving steadily. If the metal ion concentrations are kept below 1 part per billion (ppb) during the glass refining process, the losses from these impurities will be less than 1 dB/km. Modern fabrication techniques can reduce these levels to less than 0.1 ppb. Unfortunately, the non-metallic dopants needed to control the core and cladding indices of refraction act in the opposite direction and slightly increase absorption.

With metal impurities under control, the next contaminant of significance that needed to be controlled was water vapor, in the form of the OH⁻ ion. The fundamental water vapor absorption (resonance) takes place at 2730 nm, but significant overtones occur at 1390 nm and 1240 nm. Recall that as wavelength decreases, frequency increases. Notice the two water peaks in the spectral attenuation plot in Figure 3-1. Early commercial long wavelength, single-mode systems were designed for the two regions or windows which flank the 1390 nm peak. Concentrations of water vapor of less than 1 ppb can now be achieved by drying the glass in Chlorine gas during manufacture. This has resulted in much lower water peaks — and the ability to use the total spectrum from 1300 nm to 1600 nanometers for single-mode, long distance systems (Chapters 7, 8).

3.2.2 *Scattering Losses*

Light propagating in a fiber can be converted to unbound (radiative), and/or backscattered light. No matter how good the quality of the glass it will still contain small, molecular level irregularities, either within the glass or on the surface. The resulting minute variations in the glass's refractive index cause scattering. This attenuation phenomenon is called Rayleigh scattering, and its effect is shown in Figure 3-1. Core propagation will be only slightly affected by these irregularities as long as the wavelength of the light is large compared to the size of the imperfections. The scattering takes place in all directions. The scattered light represents a small loss of signal, but can still interfere with the performance of advanced high-speed systems. The energy of the scattered light is proportional to $1/\lambda^4$.

An important use of Rayleigh scattering is found in the *Optical Time Domain Reflectometer (OTDR)*. In the OTDR, light energy sent by a transmitter will be reflected back by reflections and Rayleigh scattering. The reflections help locate fiber link imperfections and the scattered light gives a measure of overall fiber attenuation. The OTDR is discussed more thoroughly in Chapter 7.

3.2.3 *Bending Losses*

Bends in a fiber can lead to increased loss. Large scale bends have radii greater than the diameter of the fiber and are called *macrobends*. Macrobends are caused by the pulling, squeezing, or bending of the fiber/cable during spooling or during installation. They can generally be seen from outside the cable. Macroscopic fiber bends of 10 cm or more will cause very little bending loss. If a 125 micron diameter fiber is bent to a radius of a few centimeters, however, there will be bending losses as well as the possibility of extreme stress and breakage. The stress and possible damage may not show immediately, which is the reason for exercising care during installation.

Microbends are a continuous succession of very small bends resulting from a number of causes: non-uniformity at the cladding/coating interface; non-uniform lateral pressures from the cabling process: microscopic variations in the location of the core axis. These cannot be seen from outside the cable. Their equivalent radii of curvature are a few millimeters. Microbend losses arise as the fiber is slightly distorted when undergoing spooling or installation. Light rays traveling near the critical angle will lose energy at the irregular interface between the cladding and buffer. In multimode fibers energy will also be coupled, and lost, from the guided modes to the leaky or non-guided modes. These losses can be minimized in fiber/cable manufacturing. Microbend losses are less serious in a loose tube constructed cable, or in a tight-buffered fiber that has a soft, flexible external covering (Chapter 2).

3.3 Dispersion (Arrival Time Distortion)

The other main parameter determining optical fiber system performance is arrival time distortion at the detector. The components in a fiber system each contribute to dispersing or spreading out the signal energy over time as it travels. *Dispersion* is defined as a spreading in time of arrival of a received signal beyond its original time spread. A system designed according to time of arrival distortions caused by dispersion is referred to as being *bandwidth* or *rise time limited*. In Chapter 7 dispersion will be directly related to available bandwidth/pulse rise time and, hence, total system performance. The fiber will generally be the greatest contributor to total system dispersion.

In the current generation of optical systems, where only intensity (power) is modulated and detected, the effects of in-band phase distortion can be neglected. Phase distortion will cause non-constant group velocity, however, which results in arrival time distortion. Consider, for example, a pulse train generated using a laser and transmitted on a fiber with arrival time distortion. Figure 3-2A illustrates what happens as pulses spread, interfere with each other, and lose amplitude as a result of the spread in arrival times of the constituent wavelengths.. The spreading has the same effect on the signal as in-band amplitude distortions have on radio signals; the baseband signal spectrum suffers roll-off. This interference and pulse distortion can be significant in some fiber systems, particularly those with long lengths of fiber. The parameter $\Delta\tau$ is used to describe the time spread beyond the undistorted time of arrival of the input signal.

Figure 3-2B illustrates how dispersion results in pulse spreading. At the time of the initial pulse each of the source's wavelengths will be modulated by the input signal. Some wavelengths will carry decreased power because the source has unequal power distributed across its output spectrum. At a later time, after travel down the fiber, λ_2 arrives ahead of λ_1. The pulse has been spread by the amount $\Delta\tau = \tau_2 - \tau_1$. A pulse's width is described by its Full-Width-Half-Max (FWHM) values in time, τ_1 and τ_2 for the initial and spread pulses respectively.

Note that this diminished received pulse amplitude is caused by dispersion alone, not by in-band amplitude distortions. In the next Section, $\Delta\tau$ will be directly related to a fiber's 3 dB bandwidth. Since the effects of dispersion on a signal are dependent on fiber length, an important measure of a fiber's quality is the bandwidth-distance product. A fiber with a specified bandwidth-distance product of 500 MHz-km, for example, would have a useable bandwidth of 500 MHz over a one km distance and 250 MHz over a two km distance. The specified frequency is at the 3 dB rolloff point on the amplitude response curve. The type of transmitted signal would dictate the required bandwidth, and thus the maximum distance available from a

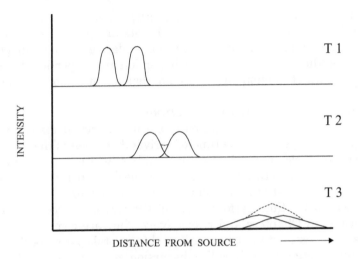

Figure 3-2A *Pulse spreading with distance.*

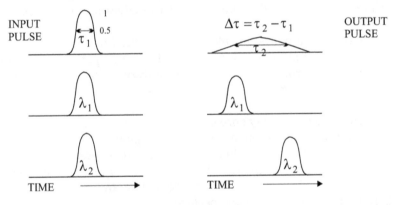

Figure 3-2B *Pulse spreading with wavelength.*

given fiber. The bandwidth-distance product of single-mode fibers is much higher than that of multimode fibers.

In a multimode fiber many hundreds of modes will be present when the signal is propagating in the steady state. The dominant cause of arrival time distortion in multimode fibers is *modal dispersion* (sometimes called *intermodal dispersion*), which is the spread in arrival times of these modes. Single-mode fibers propagate only the fundamental mode and do not have modal dispersion. Single-mode fibers have two other forms of dispersion that become dominant, however, material and waveguide dispersion. When combined they are called *chromatic dispersion* (sometimes called *intramodal dispersion*). These two contributors are present in multimode fibers, but usually make minor contributions relative to modal dispersion. The effects on the signal of chromatic dispersion are proportional to source linewidth. Modal dispersion effects are not dependent on source linewidth.

3.3.1 Delay, Bandwidth, and Risetime

This section discusses the impact of time-of-arrival distortions ($\Delta\tau$) caused by fiber dispersion. The time-of-arrival distortion is then related to the more useful measures of fiber equivalent bandwidth and rise time. Rise time and bandwidth estimates are interchangeable when components are assumed to have a first-order rolloff response at high baseband frequencies. The bandwidth/rise time performances of the optical source and detector are quantified in Chapters 4, 5 respectively. They are combined with the fiber's rise time estimate in Chapter 7 to obtain total system performance.

The next Section quantifies the dispersion generated in the two main types of silica fibers. It might appear that this sequence puts the cart before the horse, but first understanding how pulse spreading relates to bandwidth and rise time will make it easier to understand the impact of fiber dispersion.

The three main system elements (source, fiber, detector) each contribute to time-of-arrival distortion. The fiber will be the largest contributor. A four-step process is followed in analyzing the effect of time delay distortions on the signal:

- Estimate the total delay spread, $\Delta\tau$, in the arrival time of the constituent wavelengths in the intensity modulated signal
- Derive the 3 dB fiber bandwidth from this time spread
- Derive an estimate of the rise time of the fiber from the fiber's bandwidth
- Combine the fiber rise time with the source and detector rise times to derive total system rise-time performance.

The fourth step is covered in Chapter 7. Although rise times are associated with digital system pulses, this approach is also useful in estimating analog system performance.

Another approach, used when a more exact analysis of digital system performance might be needed, considers the root mean square (RMS) spreading of an impulse. The fiber's bandwidth is determined from the Fourier transform of the fiber impulse response (the $H(\omega)$ transfer function in Equation 3.1). The source and detector will also have impulse responses. Their impulse responses will be either measured or assumed. Because the source, fiber, and detector are initially assumed to have linear transfer functions, their impulse responses can be cascaded. The resulting total pulse spread relative to the input pulse will give the system performance. In both the delay time approach used in this book and the impulse response approach, the time spreading contributions of the three major system components are combined on a root-sum-square (rss) basis when determining total system performance.

Returning to the delay distortion or time-of-arrival-spread approach, the first step is to find the delay difference, $\Delta\tau$, caused by fiber dispersion. Note that the leading and trailing edges of the input pulse in Figure 3-2B are spread in their time of arrival by $\Delta\tau$. All wavelengths within the source linewidth are transmitting at the time of occurrence of the leading and trailing pulse edges. The differing delays between the wavelengths caused by dispersion will cause the arrival times of these edges to spread. The same spreading happens throughout the pulse, of course, but focusing on the edges gives a clearer understanding of the phenomena. This time-of-arrival analysis has to assume that the source spectrum is constant. This is a reasonable assumption when analyzing the effects of fiber dispersion, even though on an instantaneous basis optical semiconductor sources are not stable. In particular, at turn-on and turn-off they present unique noise problems called mode partitioning and chirp (Chapter 4).

The fiber's delay difference, $\Delta\tau$, must first be translated into a 3 dB rolloff frequency. Consider a simple sine wave modulating an optical source. The source has a wide range of wavelengths in its linewidth. As a result, the sine wave modulates all wavelengths at the same instant. Two wavelengths, one on the spectrum high side and the other on the low side, will arrive at the receiver at different times because of fiber dispersion. At some delay, these two small contributors will be detected out of phase with each other. If the delay difference between the wavelengths is equal to ½ the modulating frequency's period the contribution from these two wavelengths to the demodulated signal will cancel. This results in a rolloff in the bandwidth performance of the fiber. The frequency corresponding to this delay difference is used as an approximation of the fiber's 3 dB bandwidth:

$$f_{3-dB}(optical) = \frac{1}{T_{3-dB}(optical)} = \frac{1}{2 \times \Delta\tau} \qquad 3.2$$

where $\Delta\tau$ is the delay between the highest and lowest wavelength in the source's linewidth.

In electronic communications, bandwidth is stated in terms of a 3 dB or half power point on a curve that plots amplitude vertically and frequency or wavelength horizontally. The degree of rolloff at the high frequencies can be steep or gradual. The most gradual rolloff is caused by a *first-order response*. A first-order response in a component is sometimes modeled by a resistor/capacitor (RC) low pass filter combination. For simplicity in this book's analyses, first-order responses will be used for all components where high-frequency bandwidth shape has to be considered.

Sources, fibers, and detectors all have ½ power bandwidths in their baseband responses. The optical detector converts optical input power directly into an electrical output current that replicates the baseband signal input to the transmitter. At the fiber's 3 dB optical frequency the demodulated signal will be down 6 dB since the power in the detector output depends on the square of the current. As a result, a fiber's electrical or baseband 3 dB frequency and its optical 3 dB frequency are related by:

$$f_{3-dB}(electrical) = \frac{f_{3-dB}(optical)}{\sqrt{2}} = 0.707 \, f_{3-dB}(optical) \qquad 3.3$$

In fibers, the optical bandwidth is greater than the electrical bandwidth that is realized after demodulation. Manufacturer's specifications should be read carefully to determine which bandwidth is being presented. Usually the bandwidth given in fiber bandwidth-distance product specifications is the larger optical bandwidth. The detector alone will have a specified risetime that generally includes the effects of its associated circuitry.

A system or component that is modeled as first-order will have the following relationship between electrical bandwidth and rise time:

$$f_{3-dB}(electrical) = \frac{0.35}{t_r} \qquad 3.4$$

For purposes of simplifying the relationships between bandwidths and risetimes, the fiber is assumed to have this type of response. For a NRZ coded digital system (Chapter 7), and assuming that the received pulses have an (equal) exponential rise and fall caused by this first-order rolloff, the time delay relates to the bandwidth and rise/fall times by:

$$\Delta\tau = \frac{1}{2 \times f_{3-dB}(optical)} = \frac{1}{2 \times \sqrt{2} \times f_{3-dB}(electrical)} = t_r \qquad 3.5$$

A non-return-to-zero (NRZ) coded signal uses a pulse length equal to the bit-rate.

As an example in applying these relationships, assume a GMM fiber with a 200 MHz-km (optical) performance specification. In a 2 km link the delay spread/rise time contribution from the fiber alone for NRZ pulses would be:

$$t_r = \Delta\tau = \frac{1}{2 \times \frac{200Mhz - km}{2km}} = 5ns$$

A source will have its risetime and/or 3 dB baseband response frequency specified by the manufacturer. If only one is given, the other can be derived by assuming a first-order response and applying Equation 3.4. Usually the specification will include the bandwidth limitations of the electronic drive circuitry packaged with the source.

The time delay distortions caused by chromatic dispersion in the fiber are a function of the source's linewidth. The time average output spectrum of an optical source can be modeled as a central wavelength with ½ power or 3 dB wavelengths on either side. The wavelength difference between these two ½ power wavelengths is specified as the source's linewidth. The term *Full-Width-Half-Max (FWHM)* is used to describe this ½ power bandwidth. In addition to the source's linewidth, the average center wavelength and power must also be specified. All of these are needed in analyzing the performance of a system.

3.3.2 *Modal Dispersion,* $\Delta\tau_{mod}$

In multimode fibers the spread in energy arrival times is caused primarily by the differing modal propagation speeds (Chapter 2). This type of time distortion is referred to as modal dispersion, also called intermodal dispersion. The modes travel with different angles of ray incidence, which results in different path lengths and arrival times at the optical detector. A GMM fiber has better modal dispersion performance than a step-index fiber because the range of modal speeds is decreased.

Each mode in a multimode fiber travels at its own average group velocity. In a step-index fiber the ray traveling directly down the fiber's core will arrive at the detector in the time:

$$\tau = \frac{L}{v}, \text{ where } L = \text{fiber length, } v = \text{core velocity} = \frac{c}{\eta_1} \qquad 3.6$$

A meridional ray, describing the highest order propagating mode group, will travel at the critical angle, ϕ_c, and arrive at a greater time:

$$\tau_c = \frac{L}{v\,(\sin\phi_c)} \qquad 3.7$$

The total modal pulse spread for the step-index, multimode fiber can then be estimated at:

$$\Delta\tau_{mod} = \tau_c - \tau = \frac{L}{v\,\sin\phi_c} - \frac{L}{v} = \frac{L}{v}\left[\frac{1}{\sin\phi_c} - 1\right] = \frac{L\eta_1\Delta}{c} = \frac{(NA^2)L}{2\eta_1 c} \qquad 3.8$$

Where, from Equation 2.2:

$$\Delta = \frac{\eta_1 - \eta_2}{\eta_1} \approx \frac{\eta_1 - \eta_2}{\eta_2} = \text{Fractional index of refraction}$$

The derivation above is one of the simplest used to arrive at step-index modal spread, but is also one of the easiest to visualize. A more accurate analysis would use the effective indices of refraction for the propagating mode groups instead of the core and cladding indices, but the results would be quite similar. In an application of Equation 3.8, assume a step-index fiber with $\eta_1 = 1.5$, $NA = 0.173m$, $\Delta = 0.007$. The modal time spread per unit length becomes:

$$\frac{\Delta\tau_{mod}}{L} = \text{step - index modal time spread per unit length} = 34 \ ns/km$$

This seems like a small delay, but consider transmitting a 20 Mb/s RZ signal on this fiber. In return-to-zero coding each pulse is generally ½ of a bit period. Pulses at the fiber input are separated by 50 ns but after about 1 km the leading and trailing edges can be viewed as spreading 34 ns, thus resulting in mutual interference. Pulses would overlap and could not be reliably detected, as pictured in Figure 3-2A. Maximum signaling rates based solely on fiber-induced delay distortion are discussed later in this Chapter. Maximum signaling rates based on total system delay distortion are discussed in Chapter 7.

In graded index (GRIN) multimode fibers, GMM, higher order rays (modes) are bent gradually back to the axis as they progress down the fiber. During a large portion of their travel time, they are close to the cladding in a region of low refractive index and higher velocity. As a result, in GMM fibers the average velocity for these modes will be higher and closer to that of the axial, fundamental mode. Another way to view this effect is that rays crossing the axis at an angle at the same time will tend to arrive at the next axial zero crossing with approximately the same delay and hence suffer less travel time distortion. The spread in modal arrival times is about 100 to 1000 times better in a GRIN fiber compared to a step-index multimode fiber.

For a GMM fiber with a core refractive index shape that is parabolic (Equation 2.17), the theoretical modal time spread per unit length is given by (without derivation):

$$\frac{\Delta\tau_{mod}}{L} = \frac{\eta_1 \Delta^2}{8c} \qquad\qquad 3.9$$

Using the same parameter values as in the step-index example above, the modal time spread per unit length would be:

$$\frac{\Delta\tau_{mod}}{L} = 30.6 \ ps/km$$

The fractional index, Δ, in a GMM fiber is derived using the refractive index at the axis relative to the cladding index. Note that the modal dispersion in this GMM fiber is about 1000 times better than that of the step-index fiber example.

3.3.3 *Chromatic Dispersion,* $\Delta\tau_{chr}$

Single-mode fibers do not have modal dispersion but still have arrival time distortions. From Chapter 2, the index of refraction of the material in the glass is not a constant for all wavelengths. Since the group velocity, which is the velocity of energy propagation in the fiber, is a function of the index, the different wavelengths in an optical signal will travel at different velocities. Optical sources for low-speed/low-bandwidth fiber systems are essentially non-coherent and, hence, generate a relatively large range of wavelengths in their linewidths.

Consider the intensity of a laser pulse that is transmitted down a glass fiber in the fundamental mode alone. Different wavelengths in the pulse will not arrive at the same time, which leads to chromatic dispersion (also called intramodal dispersion). Chromatic dispersion is made up of two components, *material dispersion* and *waveguide dispersion*. Material dispersion is caused by the variation of the core's index of refraction with wavelength. As might be suspected, an expected amount of material dispersion has to be accepted once the core doping has been determined. Waveguide dispersion, which is the smaller of the two arrival time distortions, is caused by variations in the profile or cross-section of the core's index of refraction. Sometimes the cladding profile is also varied to help control waveguide dispersion. Chromatic dispersion is controlled in single-mode fibers by using the waveguide core/cladding profile to control the waveguide dispersion. The material dispersion component can be effectively cancelled at one or more wavelengths. The two components are actually somewhat interrelated, but analyzing them separately and adding their effects gives a good approximation to the total amount of chromatic dispersion:

$$D = \text{Total Chromatic Dispersion in } \frac{ps}{nm \cdot km} = D_{mat} + D_{wag} \qquad 3.10$$

In this discussion, chromatic dispersion will be calculated only for the fundamental mode, even though each mode group (in a multimode fiber) experiences some chromatic dispersion. In multimode fibers modal dispersion predominates. Analyzing the combined effects is too complicated for this book. Further, recall that the applications of optical fibers in communications today have separated into two main areas: short distance, medium bandwidth local networks using multimode fibers, and long distance, high capacity networks using single-mode fibers. If the system designs are dispersion limited, the former will be controlled by modal dispersion and the latter will be controlled by the fundamental mode's chromatic dispersion.

An exception occurs when a GRIN high bandwidth multimode fiber is used at 800 nm with a wideband (LED) source. In this case the chromatic dispersion becomes an important consideration because the source has a broad linewidth and chromatic dispersion is high in silica at 800 nm. An approximation to the combined effect of modal and chromatic dispersion is derived for this case by adding their time spreads on an rss basis.

Material dispersion occurs because the index of refraction of Silica is not constant, or a linear function of optical wavelength. Figure 3-3 shows the index of refraction of Silica and the group index/velocity as a function of wavelength. Note that the first-order derivative of the material index is zero and the group velocity is flat around 1300 nm. In this region, group velocity and material dispersion have little dependency on wavelength. Also note in Figure 3-3 how the bulk index curve is shifted by the addition of 13.5% Germanium doping. This illustrates one of the two main ways chromatic dispersion is controlled, the other being by the control of waveguide dispersion through the careful design of the refractive index profile.

The group velocity, v_g, is the speed of propagation of energy in the fiber. The propagating signal in the fiber will experience a corresponding effective group index of refraction, η_g, derived from the group velocity. The group velocity is equal to:

$$v_g = \frac{c}{\eta_g} \qquad\qquad 3.11$$

Where, without detailed derivation:

$$\eta_g = \eta_1 - \lambda \frac{d\eta_1}{d\lambda} \qquad\qquad 3.12$$

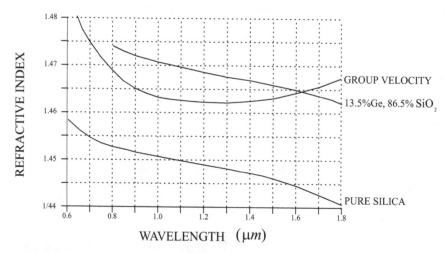

Figure 3-3 *Indices of refraction (silica).*

A more exact analysis would use the modal effective index of refraction instead of the core index, but the approach used here will suffice, particularly since it will be applied exclusively to the analysis of single-mode fibers.

The average arrival time, τ, of light transmitted over a length, L, of fiber is given by:

$$\tau = \frac{L}{v_g} = \frac{L}{c}\left(\eta_1(\lambda) - \frac{\lambda d\eta_1(\lambda)}{d\lambda}\right) \qquad 3.13$$

The pulse time spread, $\Delta\tau_{mat}$, due to a source having a linewidth, $\Delta\lambda$, then becomes:

$$\frac{\Delta\tau_{mat}}{\Delta\lambda} = \frac{d\tau}{d\lambda} = -\frac{L\lambda}{c}\frac{d^2\eta_1}{\delta\lambda^2} = -LD_{mat} \quad \text{where} \qquad 3.14$$

$$D_{mat} = \text{material dispersion coefficient} = \frac{\lambda}{c}\left(\frac{d^2\eta_1}{d\lambda^2}\right) \text{ in } \frac{ps}{nm \cdot km} \qquad 3\text{-}15$$

The material dispersion coefficient is plotted in Figure 3-4 for pure Silica and Silica doped with 13.5% Germanium. Material dispersion is large at 800 nm, and drops to approximately zero in the region of 1200 to 1400 nm, depending on the Ge content. The zero point is determined by where the second order derivative of the refractive index in Equation 3.14 becomes zero. In the 1550 nm window the material dispersion is negative,

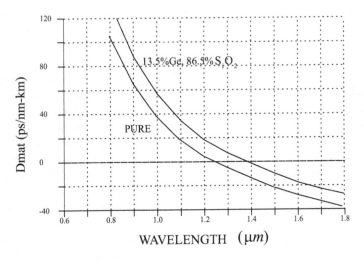

Figure 3-4 *Material dispersion coefficients (silica).*

or opposite in sign. The change in sign means that on the left side of the zero wavelength the (lower) wavelengths in a signal linewidth will be delayed more than the higher wavelengths. On the right side of the zero crossing the situation is reversed, the higher wavelengths will be delayed more. For the discussion here, the differences in delay times are the important factor, not the absolute relationship between wavelength delays. Soliton transmission, discussed later in Chapter 7, will make use of these relative delays. In some presentations of material dispersion the curves will be reversed because the minus sign in Equation 3.14 is dropped.

The effects of chromatic dispersion on a transmitted signal are directly dependent on the center wavelength of the signal and the linewidth of the source. At first window wavelengths around 800 nm, material dispersion dominates and the contribution of waveguide dispersion is small and usually neglected. From 1300 to 1550 nm the two are equally important. The amount and shape of the waveguide dispersion curve is controlled by the size and shape of the core, the smaller the core the greater the effect. This control allows the optimized single-mode fiber designs described in Chapter 2. These optimized core designs have become important in realizing Terabit/sec transmission systems (Chapter 8).

Waveguide dispersion depends on the ratio of the fiber's core radius to the signal wavelength. A detailed mathematical analysis of this dispersion component is beyond our intent, but it is constructive to examine what can be achieved through its control. The solid lines in Figure 3–5 plot the waveguide dispersion factor, D_{wag}, for a step-index single mode fiber with different core radii. The so-called conventional SM fiber has a core radius, a, of 4.5 μm, 4% Ge core doping and a pure Silica cladding. The Germanium doping plus the waveguide dispersion shifts the zero dispersion wavelength from 1270 for pure Silica to 1308 nm. Note how the waveguide dispersion component increases with decreased core radius. If the radius is reduced further to 1.8 μm, the zero would be shifted all the way out to 1750 nm, but this would make the mode field very small and hurt coupling at splices and connectors. A related parameter, the naturalized frequency V (Chapter 2), must be kept between 1.5 and 2.4 to keep the fundamental mode energy bound to the core. Recall that the cutoff wavelength occurs when V equals 2.4. Since the ratio a/λ is a prime determinant of V, reducing the radius reduces V which results in increased susceptibility to higher order mode conversion at discontinuities because of increased cladding energy.

The dotted lines in Figure 3-5 show the total chromatic dispersion, $D_{mat} + D_{wag}$, for three single-mode fiber types: conventional step-index, dispersion shifted, and dispersion flattened. The curves represent general results, but they indicate that a great deal of chromatic dispersion control is possible by controlling the waveguide dispersion component. Control is achieved through variation of indices of refraction, depth and width of cladding par-

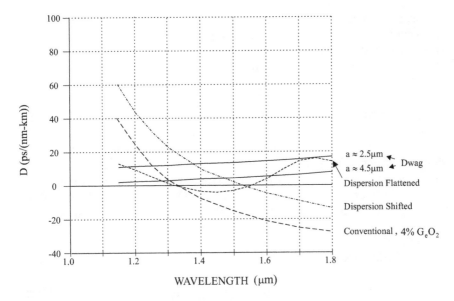

Figure 3-5 *Total dispersion-single mode fibers.*

titions, core shape, and doping concentrations. This control of the wave-guide dispersion factor, coupled with constantly improving manufacturing techniques, has resulted in the many applications of single-mode fiber seen today. Achieving low dispersion over a wide band translates directly into high system capacities and long distances between regenerators.

With the dispersion-shifted fiber, the objective is to realize a dispersion zero in the 1550 nm region. The core profile in Figure 2-12 shows that this is achieved by using the combination of lower Ge core doping, a depressed cladding using Fluorine doping, and a trapezoidal shaped core. The lower Ge content helps reduce attenuation and the trapezoidal shape helps confine the mode field. In general, these fibers have core radii between 2 and 2.5 μm, core doping of up to 13% Ge, and index differences between 0.6 and 1.8 %.

Finally, a representative curve for a dispersion-flattened fiber is plotted. An index profile for this fiber is also shown in Figure 2-12. Because of the profile this fiber is also called double or "W" cladding. Trying to use step-index designs to achieve flattening leads to unacceptably high attenuations because the Ge doping is excessive. Figure 3-5 illustrates the degree of control of the chromatic dispersion characteristics of single-mode fibers. Building on Equation 3-14, we can now state $\Delta\tau_{chr}$ for a single-mode fiber:

$$\Delta\tau_{chr} = \Delta\lambda(L(D_{mat} + D_{wag})) = -\Delta\lambda L D \qquad 3.16$$

As an example of the application of Equation 3.16, consider a laser diode centered at 1550 nm with a linewidth of 0.1 nm transmitting a signal over 1 km of the conventional SM fiber whose dispersion is plotted in Figure 3-5. The D value is approximately –17 ps/(nm-km) which results in:

$$\Delta\tau_{chr} = 0.1(1)(-17(10^{-12}) = -1.7 \; ps\,/\,km$$

Comparing this with the result of 30 ps/km for the GMM fiber example shows the single-mode fiber to be about 20 times better in arrival time distortion. If a dispersion-flattened or shifted fiber had been chosen, the single-mode fiber would be even better. Note also from Section 3.2 that a single-mode fiber will have slightly better attenuation performance than a multimode fiber. The significance of these improvements in terms of faster bit rates and longer distance transmission is discussed in Chapters 7, 8.

3.4   Polarization

Single-mode fibers have improved significantly in quality, uniformity and bandwidth-distance performance through the last two decades. The chromatic dispersion improvements discussed in the preceding section now allow greatly increased system lengths and capacities. A problem that emerges as these fibers are pushed to their performance limits is polarization mode dispersion (PMD). Depending on the quality and care in the manufacturing and cabling stages, this arrival time distortion can vary from less than 0.05 ps/$\sqrt{km}$ to several ns/$\sqrt{km}$ in conventional single-mode fiber. PMD is not a problem in multimode fiber.

From Equation 3.2, considering polarization mode dispersion only, PMD would allow a usable bandwidth for intensity-modulated systems of 50 GHz over 200 km when it is low, but only 500 MHz/km when it is high. Current systems are approaching 40 Gb/s speeds and therefore require high quality fiber to keep polarization mode dispersion low.

In Chapter 2 it was pointed out that each propagating mode in a nominally circular optical fiber actually has two polarizations. Recall that an electromagnetic wave's polarization is determined by the linear direction of its E-field, which may be stable or changing. The energy in the two polarizations becomes coupled in conventional fiber because of the fiber's unavoidable imperfections and strains. The better the quality of the fiber and cable, the closer together the two modes are to having equal transmission characteristics, and the lower the value of PMD dispersion.

PMD in conventional fibers results in an elliptically polarized wave at the detector that changes orientation and amplitude with wavelength and time. This ellipticity is caused by the two orthogonally polarized modes having different propagation characteristics and, hence, group velocities. Differential transmission for two orthogonally polarized modes is called *birefringence*. With intensity-modulated systems this causes dispersion. In coherent systems, where the receiver must have an incoming carrier with known polarization, the signal cannot be detected. Examples of coherent optical systems or sub-systems are optical gyroscopes/sensors, integrated optical circuits, and coherent transmission systems. These are all discussed in later chapters. Polarization can be maintained and used for these applications only in single-mode fibers..

Special fibers, called *polarization maintaining fibers (PMF)*, can be constructed so that their internal propagation characteristics either decouple the two modes or suppress one of them. In a linearly birefringent fiber, energy with an E-field orientation in one of the directions is not coupled into the other, and a signal with one of the polarizations will transfer very little energy into the other polarization. In a truly single-polarization fiber one of the two modes is attenuated. Both of these PMF fibers preserve the polarization of the launched signal at the detector. If an arbitrarily polarized signal is input, only one polarization will immerge.

Both types of PMF fibers are manufactured by either of two methods: inducing asymmetric internal stresses, or changing the internal geometry of the fiber through added layers of differently doped glass. Figure 3-6 gives profiles of fibers intended to control birefringence by these two methods. These fibers would be used only for signals that require maintenance of a state of polarization. They would not be used for regular intensity modulated signals because the PMD would be significant. Figure 3-6A shows a fiber with an elliptically deformed core. By deliberately destroying the circular symmetry through built in stress on the core, the two polarizations will have decreased coupling. Boron doping is used to reduce the refractive index of Silica. The different materials have different thermal expansion characteristics so that stresses are left in the glass after cooling. One polarization will stay relatively independent of the other. It is also possible to build this type of fiber by stressing the cladding while keeping the core circular.

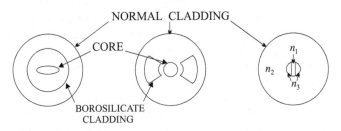

Figure 3-6 *Linear-polarization-maintaining (single-mode silica).*

Figure 3-6B illustrates a "Panda" or "bow-tie" fiber also constructed by introducing high stress regions during manufacture. This fiber construction can be used to make either a high birefringent or single-polarization fiber. One advantage of this approach is that attenuation is lower if the stressed region is displaced from the core.

Figure 3–6C shows the cross section of a single-polarization fiber. This is called a *side-tunnel* or *side-pit fiber*. This fiber's core has asymmetrical indices of refraction; a core of 1.47, a cladding of 1.46, and the side-pit of 1.45. Energy is suppressed in one of the modes because its cutoff wavelength has been separated from the main mode. Remember that in a circularly symmetric fiber, theoretically, the fundamental mode (and its orthogonal twin) do not have a cutoff.

3.5   Non-Silica Fibers

Silica fiber manufacturing and use have matured significantly. Silica fibers now are very economical and give excellent performance in many different applications. As a result, there is not a great deal of emphasis being placed on finding alternative fiber types for telecommunications applications. This section briefly discusses the general direction of research on non-Silica glass fibers, and what advances might be made in the future. Also, plastic optical fibers (POF) are discussed in some detail because they have unique properties that Silica cannot match. A combination of Silica and plastic has found use in plastic coated Silica fibers (PCS).

Research continues on materials and waveguide structures that offer the prospect of extremely low losses in the mid-infrared region (2–5 microns). We think of "glass" as the common window type glass we look through, and which also yields Silica fibers when purified. A promising type of glass for low-loss fibers is made primarily of heavy metal fluorides and zirconium fluoride. The basic challenge is to find a manufacturing approach that overcomes absorption and scattering. Also, long lengths of fibers with controlled profiles, structure and cabling must be realized. Dispersion can be controlled to levels less than 1 ps/nm-km. Probably the biggest barrier to their use lies in the lack of inexpensive components at IR wavelengths.

Plastic optical fibers (POF) are truly optical in the human sense of the word. They have their lowest attenuation, and hence most of their application, in the visible spectrum around 600 nm. The core is generally large, step-index, and made of *polymethyl methacrylate* (*PMMA*). A typical core size is 1000 microns. All-plastic fibers are multimode because of their large cores. The cladding is also plastic but not as thick on a relative basis as Silica fiber cladding. Index differences are high, leading to large numerical apertures and acceptance angles. Plastic fibers are used for short distance, very low bandwidth communications, as light pipes, and to transfer images. The automobile industry uses plastic fiber for the latter two applications, primarily because of their low weight and environmental robustness. For image transmission over short distances many smaller plastic fibers are combined into a bundle. Special plastic fibers are used in physician's endoscopes because of the need to transmit the complete visible spectrum.

4

Optical Sources for Fibers

4.1   Introduction

Radio systems require electromagnetic energy sources in the frequency range of 1 MHz to 40 GHz, which must be coherent and stable in amplitude (Chapter 1). Recall that the spectrum of coherent sources appears as a single frequency with very little accompanying noise modulation. The spectrum of a noncoherent source is a band of noise, sometimes confined but usually quite broad. Because of their coherence and stability, RF sources can transmit signals through the modulation of their amplitude, phase or frequency. They also provide coherent and stable signals at receivers for subsequent demodulation.

Optical systems, on the other hand, require sources in the range of 100,000 GHz, which is in the spectrum of visible light and the infrared. Currently installed intensity-modulated optical fiber systems use *light emitting diodes* (LEDs) and *lasers* for sources. Both are fabricated from layered semiconductors. These light sources are not as coherent or stable as radio frequency oscillators. Light emitting diodes and lasers are usually modulated by varying the injected drive current. This results in an unfortunate characteristic: their performance is influenced by the energy level of the driving signal. However, LEDs provide a very good match for silica fiber systems, because of their low cost, reliability, small physical size and emission wavelengths. This chapter discusses only injection-modulated optical diodes, which are either light emitting diodes or lasers. When improved performance is needed, lasers are modulated externally to the lasing semiconductor cavity (Chapter 7).

The technologies used to obtain radio and optical signals differ significantly. Radio system carrier frequencies are generated using electronic circuits containing gain devices and lumped or distributed passive components. Together they provide the three necessary functions to sustain oscillation: gain, feedback and frequency selectivity. The signals are coupled to electrical loads that are either other electronic circuit components or antennas.

By contrast, in optical fiber systems, the semiconductor devices convert injected electrons to photons at the desired center wavelength and intensity. The optical diode's physical design is tailored to yield radiation patterns that efficiently couple light into a fiber. The diode's internal conversion process and radiation pattern, when operating with a normal drive, will be either random (light emitting diode or LED), or controlled (injection laser diode or ILD). The communication LEDs discussed here are either *surface-emitters* (SLED) or *edge-emitters* (ELED). Light emitting diodes emit a broad band of frequencies, much like that of incandescent lights. Laser diodes are light frequency oscillators, but are constructed differently than their radio counterparts. The three functions needed to sustain oscillation (gain, feedback, selectivity), are all achieved internally in the ILD. The ILD can be constructed to be single-frequency, but even then will have significant noise modulation on the central frequency, and thus be noncoherent by RF standards. The two other basic types of lasers, gas and solid state, can be designed to be very coherent and stable but their characteristics are not compatible with optical fibers systems.

Section 4.2 discusses the materials used to make injection light emitting diodes and lasers. Section 4.3 presents LEDs, and Section 4.4 shows ILDs. The design, operating characteristics and performance of each are presented using six categories: construction, modulation bandwidth, power output and modulation, spectral emission and coherence, coupling to fibers, packaging and reliability. A *superluminescent diode* (SLD), which is constructed similarly to the edge-emitting LED but has the gain of the ILD, is briefly discussed in Section 4.3. The optical amplification necessary to sustain oscillation in ILDs, can also be used for signal amplification (Chapter 8).

4.2   Optical Semiconductors

Semiconductors for electronic circuit applications are made of silicon or GaAs. Semiconductors for optical photoemission applications are made of crystalline III-V materials: Al, Ga, and In from the III category of the periodic table, P, As, Sb from the V category. GaAs and GaAlAs devices are used for the 0.7 to 0.9 μm wavelength range; InP and InGaAsP devices for the longer 1.3 to 1.6 μm wavelengths. The substrates used for fabrication are GaAs for the 0.7 to 0.9 range and InP for the 1.3 to 1.6 range.

Optical semiconductors are diodes with their p-n junctions forward biased. Current is injected into the junction. The optical output can be *continuous wave* (CW) or modulated with the signal to be transmitted. Majority carrier electrons and holes flow into the junction region where they recombine giving the output light. The process of recombination distinguishes LEDs from ILDs. Recombination is spontaneous in LEDs, which results in a fairly broad optical spectrum. In ILDs recombination begins spontaneously but becomes controlled due to internal stimulation and feedback, once a threshold current is passed.

According to quantum mechanics, the atoms and molecules in semiconductor crystals have internal energy levels that are discrete or quantized. In optical sources, when an electron makes a transition from one energy level, E_1, to another energy level, E_2, light having a frequency proportional to the difference will be either absorbed or emitted. This is referred to as the *Bohr condition*:

$$f = \frac{E_1 - E_2}{h} \text{ Hz} \qquad\qquad 4.1$$

where h = Planck's constant = 6.625×10^{-34} J–s. The energy of injected electrons is transformed to light energy. Optical sources are thus E-O transducers. The semiconductor quantization levels can be thought of as bands of allowable energies occupied by electrons. These energies are described mathematically using statistics and probabilities. An electron will be either in an excited (conduction) state or a stable (valence) state. A number of factors, including rotational momentum, determine the energy level of a given electron. When electrons fall either naturally or through stimulation from the E_1 level to the E_2 level, a photon of wavelength $\lambda = c/f$ will be produced.

Spontaneous or natural emission occurs when these transformations occur randomly. The light we see every day and night from both artificial and natural sources is spontaneous radiation. LEDs emit spontaneously, but their possible energy bands are somewhat confined, which yields a band of emitted wavelengths. LEDs have much wider optical output bandwidths than ILDs because the range of allowable energies in the E_1 and E_2 bands are wider and the transitions are not controlled (stimulated). Also, LEDs generally emit less power than lasers. LEDs are slower in their response to drive signal variations than ILDs, and thus cannot transmit very high bit rates. The phasing in the wavefront of an LED's radiation is random which gives it spatial non-coherence. However, the internal active region of both LEDs and ILDs can be designed to radiate in desirable patterns. Semiconductor optical sources must have a constant supply of higher-level electrons to achieve emission that is steady and reliable. With both LEDs and lasers, this comes initially from the injected flow of electrons. Lasers, as opposed to

LEDs, have extended active regions that make internal gain and resonance/feedback possible. In lasers, the photons that are first generated spontaneously stimulate additional (in-phase) photons, thus giving internal gain and increased coherency.

The transition of an electron moving from a higher to a lower state, thereby emitting light, is called *recombination*. The electron finds a hole to occupy and releases energy. There are two basic types of semiconductor materials that experience these transitions:

- Direct-band-gap, in which the electron and hole in the conduction and valence energy bands have the same momentum value. These recombinations are the simplest and most likely to occur. The binary molecules of GaAs, InP, and InAs are the basic materials used for most of the current production of optical semiconductors. They are used to produce large, single crystals that can be sliced into wafers for substrates. Doped layers of similar materials are deposited on the substrates. These act to confine the electron/hole carriers, which then leads to the efficient generation of output photons.
- Indirect-band-gap, in which the momentum states between the conduction and valence band energies differ. To release radiation through recombination, an electron has to first be forced to give up momentum energy in the form of a phonon. After the phonon transition, a direct-band-gap recombination is possible. Phonons only cause heating of the crystalline structure. The group IV elements, silicon and germanium, are indirect-band-gap materials and are unsuitable as substrates for optical sources. The ability to integrate photonic and electronic signal processing on the same transmitter chip would be a big advantage. Silicon and germanium can efficiently absorb and transform photons to electrons and thus can be used for integrated optical detectors (Chapter 5).

The band gap energy difference in Equation 4.1 results in a photon of wavelength λ. Table 4-1 lists the wavelengths available with semiconductors made from a variety of III-V materials. Note that these are binary, ternary and quaternary materials. All are direct-gap materials. The ternary and quaternary compositions emit wavelengths in broad ranges, which is useful in meeting varying system requirements. Wavelength control is obtained through adjustment of the molecular ratios of the basic elements. For example, the ternary alloy $Ga_{0.93} Al_{0.07} As$ will have a gap of 1.51 electron-volts (eV) and emit at $\lambda = 0.82$ μm. The quaternary alloy $In_{0.74}GA_{0.26}AS_{0.56}P_{0.44}$ has an energy gap of 0.96 eV and emits at $\lambda = 1.3$ μm. If the concentrations are changed, the emitted wavelength changes. GaP devices are used in the visible region around the color red at 660nm. Plastic fiber is often used for visible light transmission (Chapter 2). The GaAs

Table 4-1 *Composition of emitters.*

MATERIAL	BAND-GAP (EV)	WAVELENGTH RANGE (μm)
Gallium Phosphide (GaP)	2.26	0.54
Gallium Arsenide (GaAs)	1.42	0.86
Indium Phosphide (InP)	1.35	0.92
Gallium Aluminum Arsenide (GaAlAs)	2.1-1.4	0.6-0.9
Indium Gallium Arsenide Phosphide (InGaAsP)	1.3-0.7	0.95-1.8

and GaAlAs devices are used for the shorter wavelength fiber window around 800 nm. These are typically fabricated on substrates of GaAs. The quaternary devices, matched to the longer wavelength windows at 1300 and 1550 nm, are fabricated on substrates of InP. All of these sources radiate in bands around the designed center wavelength, even though the binary combinations listed would seem to have only a single wavelength. Just as with electronic integrated circuits or devices, optical semiconductors are fabricated in layers. Because they are crystals, groups of atoms (even though doped) become arrayed in a pattern or lattice. This results in the material having a lattice constant. Sources are grown with layers of p or n doped alloys. The layers are grown as a single crystal with lattice constants matched between them. This results in a secure bond between layers. Without the lattice matching, stresses and strains on the internal interfaces caused by temperature changes or vibration would lead to failure. At these boundaries, a change in the band-gap energy occurs. The boundaries or junctions are called heterojunctions and the devices are called heterojunction optical semiconductors. The term *hetero* means "containing different atoms." The layers confine the recombination process to a specific controlled area, which then allows control of the diode's performance and radiation pattern. For stripe devices (ELEDs, SLDs, ILDs), the internal confinement area becomes a waveguide, much like the core of a fiber. The LEDs and ILDs discussed in the next sections are all double heterojunction devices because the active region is confined on top and bottom by a heterojunction. A simple doped p-n junction, called a *homojunction*, would radiate, but it would not be an efficient source for fiber systems.

These heterojunction devices are fabricated by epitaxial techniques. With epitaxy growth, chemicals in gas or liquid form are condensed or precipitated onto the beginning substrate in layers. There are three approaches used to obtain epitaxial growth for optical semiconductors: liquid-phase, vapor-phase and molecular-beam epitaxy. The challenge is to grow layers that are defect free, of precise thickness and of the proper molecular composition.

Electrically, these diodes have power supply demands somewhat similar to those of conventional diodes. They are forward-biased at 1 to 2 volts, instead of the typical 0.7 volts in a silicon diode. Current drains are higher, as will be shown in Sections 4.3.2 and 4.4.2. Since the current is modulated with either an analog or digital signal, the drive circuitry will be more complex. Adding to this complexity, particularly for lasers, will be the need to include temperature and current bias control circuitry.

4.3 Light-Emitting Diodes (LED)

4.3.1 *LED Construction*

Figures 4-1 and 4-2 show cross section schematics of the two main types of LEDs used for fiber systems, a *surface-emitter* (SLED) and an *edge-emitter* (ELED). Both provide high radiance, incoherent light in a relatively broad band of wavelengths. High radiance or brightness refers to the ratio of the radiated intensity in a differentially small spherical area to the corresponding emitting area on the surface of the device.

A common SLED structure is the Burrus or etched-well LED sketched in Figure 4-1. The surface-emitter radiates vertically. The ELED has a longitudinal active region in a stripe that radiates on the edge of the block. The ELED is constructed much like an ILD, but the requirements for oscillation

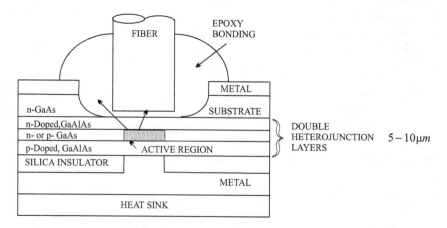

Figure 4-1 *Surface emitting LED (SLED).*

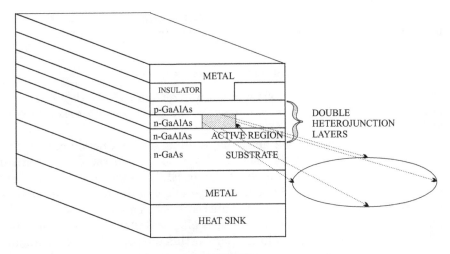

Figure 4-2 *Edge-emitting LED (ELED).*

needed to make a laser are not met in the LED. A third type of edge-emitting LED, called a *superluminescent diode* (SLD), is used for sensor applications. The SLD is constructed like an ELED. It has internal gain like a laser, but does not have a resonant cavity and controlled feedback.

The surface emitter in Figure 4-1 has a circular active area of 50 μm diameter with a thickness of a few microns. Current is directed between the two metallic layers, through the active layer. The active layer has a wide p/n junction where recombination and radiation can take place. The band gap and refractive index differences of the adjoining (double) heterojunction layers confine the recombination to the central active layer and region. This yields an efficient device with high radiance (intensity). The AlGaAs double heterostructure is approximately 50 microns thick, including substrate. A major concern regarding all optical semiconductors is keeping the active region's current density at reasonable levels to reduce internal heating and prolong the life of the device. The double hetero structure is crucial when reducing current density and achieving reasonable optical output powers. The output of an LED will drop by 2 to 3 dB if the internal temperature rises 100 degrees centigrade. Internal device efficiencies of up to 60 percent are achievable. This excludes radiation pattern and coupling losses. Internal parameters influencing efficiency and performance are: the degree of absorption of photons in the active region, recombination at the interfaces without radiation, optimum doping concentrations, and active layer thickness.

The surface emitter's emission pattern is circularly symmetric (cone-like) and has a 120° half power width. The shape is called Lambertian and is defined by

$$I = \text{Intensity (power)} = I_p \cos \theta \qquad \qquad 4.2$$

where I_p = peak power on axis and θ = angle from axis. With a Lambertian pattern, the intensity is 50 percent of the maximum on-axis intensity at $\theta = 60°$.

The ELED in Figure 4-2 has a longitudinal active junction region. The current flowing between the metallic layers is concentrated in a central stripe. The output power available from an ELED is generally a little less than that from a SLED. The active stripe is generally about 100 microns long, 50 microns wide, and about 1 micron thick. This is a good match for multimode fibers. The output intensity is elliptical if viewed in a plane parallel to the face. The pattern has a Lambertian shape in a plane parallel to the layers, but is more directional in the perpendicular plane. The half-power angles in the perpendicular plane are reduced from 60° to 30°. As a result, coupling to a multimode fiber is more efficient than with a SLED. Because of incoherency, all LEDs will generate thousands of independent spatial modes. Based on physical optics, the highest attainable coupling efficiency equals the ratio of the number of guided fiber modes to the number emitted. Lenses can be used to help achieve this theoretical maximum.

4.3.2 LED Modulation Bandwidth

The response of optical sources to a modulating signal is described in small signal and large signal responses. The small signal response is obtained by driving the device with a small amplitude sine wave, swept in frequency through the baseband. The large signal response is obtained by measuring the rise and fall times of a pulse or square wave. The two responses yield the modulation bandwidth and rise/fall time performances, respectively. The two responses are related. Through Fourier and Laplace analysis, the response of a source to a pulse is directly related to the width of its modulation bandwidth. Optical detectors used to obtain these responses must have wide bandwidths so that they do not distort the results.

Small signal analysis is particularly important for LEDs since they are often preferred for analog modulation because of their linearity. Large signal analysis is very helpful with ILDs because they are generally used for digital modulation. Also, their small signal response curves are more complex and difficult to apply in analyses.

The small signal modulation performance of LEDs can be modeled using a first-order RC type response. Parasitic capacitance will sometimes be important, but the response is generally controlled by carrier lifetimes.

If the drive current is varied at a baseband wavelength (ω), the intensity (power) of the optical output as a function of this mobility is stated as

$$I(\omega) = \frac{I_0}{\sqrt{1 + (\omega\tau_c)^2}} \qquad 4.3$$

where I_0 = emitted intensity at zero modulation frequency, τ_c = average carrier lifetime. The ac optical intensity (power) is reduced by 0.707 at $\omega = 1/\tau_c$. An optical detector's current is proportional to this intensity, so the detector's output electrical power will be down $0.707^2 = 0.5$ at this baseband frequency. This is the electrical, or modulation, 3 dB frequency for the source.

The modulation (baseband or electrical) bandwidth is less than the optical bandwidth. When the optical power is down 1.5 dB at a specific modulation frequency, the electrical power will be down 3 dB. At the optical 3 dB frequency, the electrical power is down 6 dB. The relationship between the two is

$$f_{3-dB} \ (electrical) = 0.71 \ f_{3-dB} \ (optical) \qquad 4.4$$

The manufacturer's data must be read carefully to know which bandwidth, optical or electrical, is being specified.

Both the small and large signal responses are limited by carrier recombination times. Edge-emitters have faster times (bandwidth) than surface emitters. Modulation bandwidths in excess of 100 MHz are available with ELEDs. Figure 4-8 plots and compares a typical small signal LED response with a semiconductor laser response. Note the first order rolloff for the LED.

Large signal responses are characterized by rise times. Rise times are generally defined as the time it takes for the output of a device to change from 10 to 90 percent of the final value when the input is a step function. The leading edge of a pulse is a step function if there is no distortion. The basic assumption behind this analytical approach is that the component (fiber or otherwise) responds like a first-order system. With a first-order system, the rise and fall curves at turn-on/turn-off are smooth, as seen in a capacitor's charge/ discharge. This yields a useful relationship between 3 dB electrical bandwidth and rise time given by

$$f_{3dB\text{-}electrical} = \frac{0.35}{t_r} \qquad 4.5$$

LED rise times are governed primarily by carrier mobility. Typical LED rise times range from a few to a few hundred nanoseconds. The wider the modulation bandwidth, the faster the rise time. Circuitry can be used to spike the rise and fall times, thus improving them up to 50 percent. The disadvantage is an added overshoot.

4.3.3 *LED Power Output and Modulation*

The SLED illustrated in Figure 4-1 typically will output "available" power of about 15 mW CW (nonsaturated) at a drive current of 150 mA. Available power would be that measured by a photodetector in very close proximity to the emitting surface. The LED will have a terminal voltage of about 2 volts, which gives it a device efficiency of about 5 percent. Not all of the radiated power will be propagated by a fiber because the LEDs radiating pattern is broad. A multimode fiber will be able to propagate about 5 to 10 percent of this power, depending on the fiber's numerical aperture (Chapter 2). Note that the LED in Figure 4-1 has an epoxy bond to stabilize the butt coupling of the fiber to the source. The epoxy also acts to index-match the LED surface to the fiber, thus improving efficiency a few percent by lowering reflection losses. An ELED will generally output less power but will have higher coupling efficiency because of its narrower radiation pattern.

Because a single-mode (SM) fiber has a much lower NA, it will be able to propagate less than 1 percent of the available output power. This is one of the two reasons LEDs are seldom used as sources for single-mode fibers. The other is the LEDs wide spectral bandwidth. Chromatic dispersion controls distortion in single-mode fibers. The effects of chromatic dispersion are directly dependent on the source's optical bandwidth. The wide LED spectrum results in very high signal distortion in SM fibers, except on short connections.

The output optical power of an LED is a function of the injected input current. This is illustrated in the generalized curve of Figure 4-3. The slope of the power output-current drive curve is a measure of the overall quantum efficiency of the device. A sine wave analog modulation signal is shown with about 50 percent modulation depth. Figure 4-9 shows pulse modulation with the laser diode curve. Driving the LED diode with a stronger signal than that shown would cause significant distortion in the output. The bias current for most LEDs is in the 10s of milliamps range. The region before saturation is reached has good linearity in most LEDs. Total harmonic distortion tends to be down about 40 to 50 dB from the fundamental when the index of (amplitude) modulation is 50 percent. This is satisfactory for analog video transmission over short distances. Linearity is less important for digital modulation. A small bias is often used with LED digital modulation to control the internal shunt capacitance and improve the response.

An important LED figure of merit is the power-bandwidth product, which is constant for a fixed value of injected current. If increased power output is desired, it has to come with reduced bandwidth. Conversely, faster LEDs will emit less power.

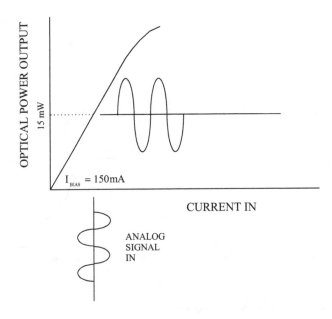

Figure 4-3 *Sled power out vs. current in.*

4.3.4   LED Spectral Emission and Coherence

The peak central wavelength of a double heterojunction LED is determined by the bandgap of the active layer. The spectral width and central wavelength are dependent on the doping concentration of the active region. Because the energy given off with each recombination is random, the range of wavelengths in the spectrum is broad. Figure 4-4 illustrates optical spectra for InGaAsP LEDs centered at 1.3 microns. Spectral width is the width of the spectra at half-power (Full Width Half Max). Notice that the edge-emitter has a narrower spectrum than the surface emitter. A GaAlAs LED operating at 0.85 microns will have FWHM spectra of about 40 nm.

The effects of chromatic dispersion on an intensity-modulated optical signal are directly dependent on the source's spectral width. Systems using multimode SI fibers will have high modal distortion. The modal distortion will outweigh the effects of spectral width dependent, chromatic dispersion. When GMM fibers are used, modal distortion is lessened and the source's spectral width could become a factor in system performance (Chapter 8). In general, LEDs are used for short distance data networking and video system applications. These generally use multimode fibers and the 0.8 micron window. Systems requiring better performance and longer distances use longer wavelength lasers and single-mode fibers.

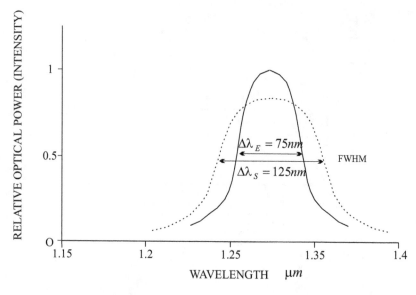

Figure 4-4 *LED optical spectrum.*

In Section 4-5 three aspects of electromagnetic wave coherence are discussed: spatial, which applies to the phase regularity in a plane normal to the direction of propagation; polarization, which applies to the E-field polarity; and temporal, which applies to the phase being predictable from one time to another at a point on the path of propagation. LEDs do not possess any of these characteristics of coherency. Spatially coherent sources can be focused with lenses to a fairly small (diffraction limited) spot, but a nonspatially coherent source cannot. Lenses are helpful in improving LED-fiber coupling only if the source has an emitting area smaller than the fiber core. Temporal coherence is directly related to the linewidth (FWHM spectral spread) of the optical source. The linewidths of LEDs are relatively broad because they emit spontaneously. As a result, LEDs have no temporal coherency and cannot be frequency/phase modulated or used for local oscillators.

The superluminescent diode (SLD) is a diode constructed similar to an edge-emitting LED but with a longer active region. It has internal gain, like a laser, but the active region is designed without feedback so as to suppress oscillations. The SLD has a spectrum about 1/10 the width of an LED. Its higher output power and narrower spectrum without coherency makes it useful in sensor applications.

4.3.5 LED Coupling to Fibers

The light that can be coupled from the source to the fiber depends on many parameters: LED emission pattern and intensity; source coherency; fiber refractive index profile and core size, fiber NA, distance, alignment, and medium between source and fiber. When the fiber core is larger than the LED emitting area, the power gathered by the fiber is approximated by

$$P_F = P_{LED}T_T(NA)^2 \qquad\qquad 4.6$$

where P_{LED} = total available power from LED, T_T = tranmissivity = 1-Fresnel losses, NA = fiber's numerical aperture.

There are two reflections that decrease coupling efficiency between the source's radiating surface and the fiber core surface. The Fresnel reflection loss at the semiconductor-to-air interface will be rather high, approximately 25 percent (Chapter 2). The air-to-fiber interface will have about a 0.2 dB (4 percent) reflection loss. For a typical silica multimode fiber with NA = 0.24, a surface-emitting LED with a total CW output power of 15 mW would be able to couple about 0.6 mW (–2.2 dBm) into the fiber for propagation.

Figure 4-1 shows that epoxy can be used between the surface emitter and the fiber to improve coupling and stabilize alignment. The epoxy glue has a refractive index close to 1.5, which provides a better optical match between the source and fiber than air. With the use of epoxy, the transmissivity rises to about 0.9 from 0.7 and the coupled power increases by about 15 percent. The most important factor in Equation 4.6 is the fiber NA (Chapter 2). A plastic fiber, for example, will have an NA about two times that of a silica fiber, which increases the coupled power by a factor of 4. Plastic fibers have significant bandwidth and transmission loss problems so this advantage is only useful in limited applications.

With the incoherent signal from an LED, significant energy is coupled into large-angle rays that begin to propagate but are converted after a short distance into cladding or so-called leaky modes. This results in lost energy after the modes travel only a few meters in the fiber. The attenuation of the fiber at the receiver will appear to be higher than expected if this is not addressed. Therefore, fiber measurements should be made only at relatively long distances from the source, i.e., after propagation conditions have been allowed to stabilize.

The edge-emitting LED has higher coupling efficiency than the surface emitter because it radiates in a more confined pattern. This advantage diminishes as the fiber's NA increases.

Because of its lack of spatial coherence, the output of an LED cannot be focused into an image of greater brightness (radiance) than that impinging on the lens. Lenses can be used to improve coupling, however, if the source has a smaller emitting area than the fiber core. Two approaches are illustrated in Figure 4-5. The lenses magnify the emitting area to match the core area of the fiber. Another way to describe this effect is that the lenses decrease the effective angle of divergence of the source, thus allowing the fiber to gather more light within its acceptance angle. Epoxy can be applied to improve and stabilize coupling. A fiber lens is made by tapering and shaping the end of the fiber (see Figure 4-5A). Sometimes fiber lenses have spherical rather than tapered ends. The spherical lens in Figure 4-5B is bonded to the LED. These small lenses are called *microlenses*. They are also separately mounted, which increases alignment difficulties during manufacture. Microlenses come in many different shapes.

When lenses are used, alignment becomes a more critical issue. The coupling efficiency achievable with lenses is based on the principles of conservation of energy and radiance. Theoretically, a 50 micron diameter, surface emitting LED (Lambertian emitter) using a lens with a 100 micron diameter fiber, NA = 0.15, will be able to increase coupling efficiency by about 3 dB. Lenses are also used to increase the distance between the source and fiber, thus easing some of the assembly and packaging problems and cost.

A single mode (SM) fiber has an NA about half that of a multimode fiber, which greatly decreases butt-coupling efficiency when the source is a large-surface LED. A surface emitter will only couple about 100 microwatts (–10 dBm) into a SM fiber. Further, lenses will not improve coupling since the source is larger than the fiber. Injection lasers have very small emitting areas and a focused radiation pattern. Their coupling to single-mode fibers can be improved by using lenses (Section 4.4.5). Edge-emitting LEDs have an emitting area smaller than an SLEDs, but their emissions

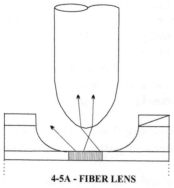

4-5A - FIBER LENS

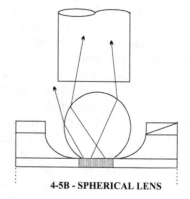

4-5B - SPHERICAL LENS

Figure 4-5 *SLED-fiberlens coupling.*

are still broad and spatially incoherent. The coupling to a SM fiber is not improved by using (conventional) lenses. A tapered fiber lens will slightly improve coupling between ELEDs and SM fibers (see Figure 4-11).

4.3.6 LED Packaging and Reliability

The different applications for LEDs require a variety of device packages. These must meet requirements in the areas of physical size and positioning, environmental sensitivity, thermal regulation, and sometimes must accommodate integrated electronic drive circuits. This section discusses stand-alone packaged LEDs. Integration of source and detectors with electronic circuitry will be discussed in Chapter 7.

Figure 4-6 shows a schematic of a packaged, hermetically sealed edge-emitting LED (TO-18 type package). The LED is attached to a metallic stud to provide electrical continuity and heat sinking. The other electrical lead is bonded to the top of the semiconductor. An asymmetric lens to magnify the output is included. The package also contains a fiber pigtail, which generally will be purchased with an attached connector. If no fiber is attached, the package would have only a glass port. The fiber would then be simply butt coupled. Control of temperature is generally important because both lifetime and power output decrease with increasing temperature. Also, temperature changes can shift the operating center wavelength. *Thermoelectric* (TE) coolers are used for devices requiring very precise control (Section 4.4.6). LEDs can be obtained in many different packages: TO, DIP, surface mount. Edge-emitting LEDs will almost always use DIP or surface mount packaging.

The lifetimes of optical semiconductor sources are affected by their fabrication techniques, operating conditions, type and quality of their packaging. Life testing of packaged sources can be carried out at room temperature or accelerated using elevated temperatures. Using accelerated

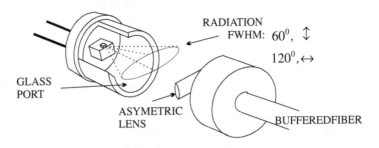

Figure 4-6 *ELED TO 18 package with lens.*

testing, one approach measures the current change at maximum CW drive before the device saturates. Another measurement approach is to find the time necessary for the output to decay 3 dB under a normal CW drive. With normal use conditions, optical output power will decrease naturally as the device ages.

Devices with lifetimes exceeding 10,000 hours (CW) are now generally available. The device's operating lifetime depends inversely on the active region's current density. For example, if the current density is doubled, the expected lifetime decreases by a factor of 3 to 4. Since optical sources are not 100 percent efficient, any excess current represents energy that turns to heat. This is the most important reason for using heat sinks. Effective heat-sinking dissipates a large amount of energy, thus lowering the junction temperature and internal stresses. Individual communication quality LEDs, with only the drive leads attached, can be purchased for a few dollars. Packaged devices with pigtails and connectors will be 10 to 100 times more expensive, depending on bandwidth, power, cooling and so on. The costs for these devices are discussed more thoroughly in Chapter 8 in the context of total system costs.

4.4   Semiconductor Lasers

The basic principles of the Ruby laser, discovered in the early 1960s, were extended to semiconductors operating at room temperature. The development of optical fiber systems was spurred by the potential availability of small, cheap, partially coherent semiconductor light sources. One very good feature of the semiconductor laser is the ability to control and modulate the injected current, thus it is also called an *injection laser diode* (ILD). Laser diodes can be constructed with very wide modulation bandwidths and fast rise times. This makes them ideally suited for digital signaling at many gigabits/sec.

The ILDs discussed here are all of the stripe geometry type. Chapter 8 discusses a new type of semiconductor laser using a vertical cavity. Vertical cavity designs offer a number of advantages over the conventional edge-emitting designs.

4.4.1   ILD Construction

This section discusses ILD construction and the modal structure of the optical wave internal to the laser. Figure 4-7 illustrates the basic construction of an index guided, edge-emitting semiconductor laser. For the past 35 years, the semiconductor lasers used for communications have all been edge-emitters. This type of construction has offered the best compro-

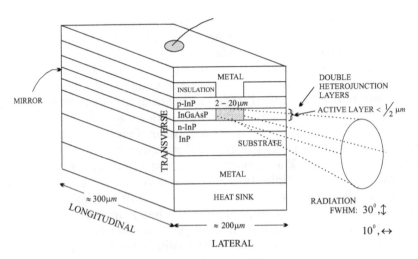

Figure 4-7 *Stripe geometry, index guided ILD.*

mise between cost, package size and performance. For communications devices, the major physical differences between the edge-emitting LED and the edge-emitting ILD are

- The active region between the confining, double heterojunction, p/n layers is thinner vertically and narrower horizontally in the ILD.
- Reflectors, sometimes with multiple layers, are added at the ends of the ILD to provide feedback and a resonance cavity.
- The ILD will have an active region length of 200 to 400 mm, which is about double the length of the ELED.
- The ILD may have a small amount of light extracted at the end opposite to the output for controlling the operating bias.
- The ILD package will have circuits to control temperature and light output.
- The ILD package will almost always be purchased with a connectorized pigtail.
- The ILD package may have optical isolators to prevent reflected energy from interfering with laser performance.

The sides of the laser in Figure 4-7 are rough-cut and not critical to performance. The ends of the cavity are cleaved along the crystalline material. This provides a good reflective surface of about 32 percent based on Equation 2.4. The resulting feedback and resonance is improved by adding reflective layers on the end opposite the output. A small amount of power is sometimes extracted at this end to provide light output monitoring and control. Any detected output power change would be translated to a compensating change in the laser's bias.

The length of the cavity determines the wavelength(s) or modes found in the output spectrum. These are referred to as longitudinal modes. Both single and multilongitudinal mode lasers are used. The former has a more complex structure and package, and is more expensive. The light stimulated in the active region will make a number of passes between the reflective surfaces, which results in internal standing waves. A standing wave indicates resonance. The resonance occurs at wavelengths determined by the cavity length. Multiple standing waves will be present in a multi-longitudinal mode ILD. A multimode ILD has significant gain over a broad range of wavelengths. A single mode ILD has gain over a narrow wavelength range, which is achieved through additional tuning designed into the active region or adjacent to it. The gain is brought about first by the drive current pumping a large number of electrons into an excited state. Then photons given off by recombination stimulate the release of more photons at the same wavelength.

The cavity in Figure 4-7 uses only the two ends to control frequency. This is called a *Fabry-Perot resonator*. The standing waves are created at wavelengths determined by

$$\lambda = \frac{2nL}{n}$$
4.7

where n = refractive index of stripe, n = positive integer.

For an example of the effect of the cavity length on output wavelength, assume an ILD designed for λ = 1.3 µm with a cavity λ = 250 µm and n = 3.5. The resulting integer n from Equation 4.7 becomes 1346. If n were to change by 1 due to, for example, a very slight change in temperature, the wavelength would change by a small amount. If $\lambda = \lambda + \Delta\lambda$ and n = n + 1, Equation 4.7 yields the difference between possible oscillating wavelengths.

$$\Delta\lambda = \frac{-\lambda^2}{2nL}$$
4.8

In the example discussed, $|\Delta\lambda| \approx 1$ nm. When the cavity length (L) is long relative to the center wavelength and the gain shape is wide, there will be multiple wavelengths setting up standing waves. These are the longitudinal or axial modes. As with any oscillator, the gain must be positive at the oscillating wavelengths. In Fabry-Perot lasers, there is positive gain over a broad spectrum. As a result, many longitudinal modes will propagate internally. The output spectrum will be broad and appear to have spikes if viewed on a spectrum analyzer.

The cross section in Figure 4-7 illustrates the internal layers of a double heterostructure Fabry-Perot ILD. Note the metallized layers that guide the injected current through the active layer's recombination region. The lattice-matched layers on all sides of the small recombination region confine the wave in the lateral and transverse directions. This laser is index-guided because the side regions have different indices of refraction relative to the active region. Because this laser uses index guiding, the recombination region becomes an optical waveguide bounded by step-index confinement layers. A gain-guided laser uses injected current density variation across the active region to achieve the same effect.

The current driving the laser is restricted to a narrow stripe along the length of the active region. Because there is wave propagation in the active region, there will be a modal pattern in the direction normal to the cavity standing wave. As with a fiber, the pattern and number of these transverse modes depends on the dimensions of the recombination region and the index of refraction differences between the active and confining layers. It is desirable to have only a fundamental mode propagate, which dictates that the recombination region be relatively small. Only a fundamental transverse mode will be able to propagate if the index of refraction difference is about 0.08 and the active area thickness held to less than a micron. In the lateral direction, a width of less than 10 microns is needed to propagate only the single fundamental mode.

The other basic type of ILD uses gain guiding to achieve the confined active stripe. With gain guiding, the metallized layers on the top and bottom focus the current more precisely. This increased current density raises the index of refraction in the central core, which then guides the wave. Index guiding gives more stable and desirable lateral/transverse mode performance than gain guiding.

4.4.2 ILD Modulation Bandwidth

Recall that the output intensity frequency response for the LED was modeled as a first order system (Equation 4.3). The modulation frequency response for an ILD is more complex. Injection lasers have a multi-pole transfer function instead of the single pole of the LED. They also have a resonance or relaxation-oscillation frequency (f_R). The magnitude of the peak response at f_R is a function of the power output, device physical parameters, and operating point. A typical ILD modulation response, with resonance, is plotted in Figure 4-8.

Figure 4-8 also includes a typical first order LED response for comparison. Note that the laser response falls sharply above f_R. This ILD has a 3 dB frequency of about 3 GHz, as shown in Figure 4-8. Because the positioning and intensity of the resonance peak depends on the average drive current, most applications depend only on the flat portion of the frequency

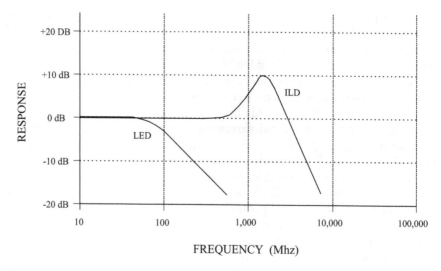

Figure 4-8 *Small signal response.*

response. This still offers a very wide frequency band for a baseband signal spectrum. As a result of this broad bandwidth, ILDs can be pulsed at many gigabits/sec; LEDs can be pulsed at maximum rates of about 200 Mbits/sec. Laser diodes can be modulated externally at bit rates up to 40 Gb/s (see Chapter 9).

The rise and fall times of lasers are much faster than those of LEDs. The wider a device's small signal bandwidth, the faster will be the response. Since the ILD does not have a single-pole response, it is not valid to relate a 3 dB bandwidth to a rise time using the LED approach in Equation 4.5. The relaxation-oscillation response in the ILD gives added bandwidth, but also results in unwanted ringing on the leading edge of a pulse. With a laser transmitter, the manufacturer's specification for rise time should be used. Caution is advised, however, because some manufacturers use 20 to 80 percent rise times instead of 10 to 90 percent. Rise times of a few picoseconds are possible with ILDs, compared to a few nanoseconds for LEDs.

4.4.3   ILD Power Output and Modulation

ILDs can be intensity-modulated using analog and digital methods. Figure 4-9 gives a typical curve of CW optical power output vs. input drive current with a superimposed digital signal. At low drive currents, the laser is emitting with spontaneous radiation, similar to an LED over its whole input range. The optical output spectrum and radiated beam-width will both be broad. As the current is increased beyond the threshold current

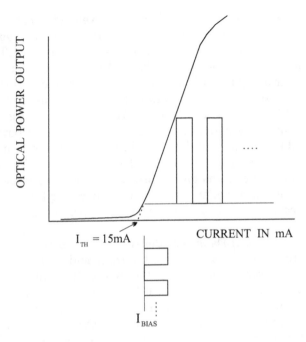

Figure 4-9 *ILD power out vs. current in.*

(I_{TH}), the gain increases and stimulated emission begins. In the lasing region beyond the threshold, the spectrum and the beam-width both narrow considerably. A single ILD can emit from a few to about 10 mW, CW. Laser arrays can be used to output more total power. As with any power output stage, optical or electrical, saturation will inevitably occur and is to be avoided.

The ILD output curve is less linear than that of the LED. If used with an analog signal, harmonic distortion will be higher. The size of the recombination portion of the active layer determines how many modes will propagate internally in the ILD. If the region is relatively large, more than just the fundamental mode will propagate. When the lateral dimension is too large, a problem known as *kinks* can develop. A kink shows up as a shifting or discontinuity in the (CW) power output curve at some specific drive current. Kinks would trouble both analog and digital signals. Index-guided ILDs are less likely to suffer from kinks because their active region can be kept smaller.

Note that the bias for the pulsed signal in Figure 4-9 is slightly above the threshold current. This avoids a phenomenon called *chirp*, which occurs when the ILD is brought quickly through the threshold by a rapid change in drive. Chirp is manifested by a changing longitudinal modal structure in the output spectrum. The cavity's resonance wavelengths change because the rapid change in drive current changes the active region's index of refraction. Sometimes the bias is set lower, i.e., on the left of the knee. The advantage here would be that in the absence of a pulse, there would be no optical output. Current drain by the ILD and receiver background noise would also be lessened. Any chirp penalty would have to be acceptable at the outset.

The threshold current is an important ILD parameter. It is defined as the current obtained by extending the slope of the lasing power output curve downward to the horizontal axis. In Figure 4-9, this is 15 mA, a value typical of index-guided ILDs. Gain-guided ILDs will have higher thresholds. It is desirable to have as low a threshold current as possible. Lower thresholds mean lower power supply demands, operating temperatures and current densities. All of these lower the cost and increase the reliability of the packaged laser.

The performance of ILDs is very dependent on temperature variation. Small temperature increases will raise the threshold current, which shifts the output power curve to the right in Figure 4-9. In addition to a threshold shift, the diode will saturate sooner. ILD communication transmitters should contain some form of temperature control to keep the threshold current and the output power curve relatively constant. The schematic of a packaged ILD in Figure 4-12 shows a thermoelectric cooler (TEC) being used to control the diode's temperature.

4.4.4 ILD Spectral Emission and Coherence

The longitudinal mode structure dictates the spectrum of an ILD. A Fabry-Perot laser will have a total linewidth of 3 to 6 nm (300 to 600 GHz). There can be up to 8 modes inside the spectrum's envelope. Gain guided ILDs will likely have multiple longitudinal and transverse modes with their FWHM width on the order of a few nanometers. Index guided ILDs will have narrower spectra and will stabilize with one or two modes as the drive is increased above threshold. Single longitudinal mode ILDs (SLM lasers) are fabricated with sharper internal tuning capability. The gain shape that determines the oscillating wavelengths is narrowed considerably compared to a Fabry-Perot laser.

Two spectral outputs, *multiple* and *single longitudinal mode,* are pictured in Figure 4-10. The spacing of the multimodes is in accordance with the Fabry-Perot cavity equation (Equation 4.7). The linewidth of an individual mode will vary from about 0.01 nm (a few GHz) in Fabry-Perot ILDs to about 10^{-4} nm (tens of MHz) in single-longitudinal mode ILDs. Because the electron/hole recombinations that take place between energy levels are determined statistically, there will always be significant noise modulation, even on a single mode. Even a SLM laser has too broad a linewidth for coherent modulation/demodulation purposes.

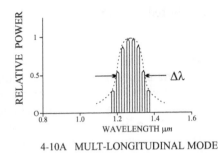

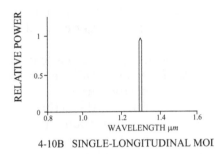

4-10A MULT-LONGITUDINAL MODE   4-10B SINGLE-LONGITUDINAL MOI

Figure 4-10 *ILD optical spectra.*

The performance of an ILD can be analyzed and predicted in detail by using *rate equations*. These equations give the electron density rate of change and the photon density rate of change in the active layer. For our discussion it is important to realize that the ILD is not a simple device with stable, repeatable performance. Many external factors can vary slightly and change performance: temperature, drive current, signal, aging, and reflections. For example, multiple modes, if present, will exchange power with each other as they grow or decay during modulation. The modal variations in an ILD's output spectrum will result in amplitude and wavelength variations. This will lead to system degradations because of the presence of dispersion in the fiber (Chapter 3). The amplitude variations lead to modal noise with multimode fibers. The wavelength variations lead to chromatic dispersion penalties. One way to control laser degradations is to operate the semiconductor laser in CW output with an external modulator.

Spectral width is directly related to a source's temporal coherence. The narrower the spectrum, the more coherent the source. ILDs have varying degrees of temporal coherence, as evidenced by the linewidths discussed above. The spatial coherence of ILDs is generally high, but best with index-guided devices. Recall that spatial coherence increases the focusing capabilities of the ILD. ILD light can be launched efficiently into both 50 micron diameter MM fibers and 10 micron diameter SM fibers.

Figure 4-10B illustrates a *single-longitudinal mode* (SLM) ILD output. To obtain a narrow SLM linewidth, the Fabry-Perot laser can be improved in various ways. These approaches act to increase the device's wavelength selectivity:

- Reduce the cavity length, which increases the spacing between modes (see Equation 4-8) so that fewer are supported by the gain curve of the active region.
- Use two coupled cavities, each detuned from the other, which yields a single mode.

- Use shortened length reflections distributed continuously throughout the lasing section or coupled to it.

The last approach has lead to a widely used SLM design called the *distributed feedback* (DFB) laser. In one popular solution, corrugations that are close together are deposited on top of the active layer. The stimulated light generated in the active layer's recombination region is continuously coupled to the distributed tuned circuit, resulting in a single-longitudinal mode output. Reflections from within the main cavity are subordinated. One of the ends might be cleaved and the other, the output end, tilted. The tilt also reduces reflections back into the cavity from the fiber that might serve to detune the oscillation. Linewidths of just a few megahertz are possible with DFB lasers. External distributed cavities, called *distributed Bragg reflector* (DBR), can also be used. Their linewidths are narrower, but they tend to have higher thresholds, more temperature sensitivity and more chirp.

4.4.5   ILD Coupling to Fibers

It is important to couple as much energy as possible from the laser into the fiber. Lasers have radiation patterns that are narrow in the plane parallel to the lateral active region and wide in the perpendicular plane. This is opposite to that of the edge-emitting LED because the laser light has spatial coherency. The ILD's radiation pattern is also narrower in both dimensions. The perpendicular FWHM emission is 30° to 50°, and the FWHM parallel emission 5° to 10°. The ILD radiation pattern is controlled by the effects of edge diffraction, which occurs only when spatial coherence is present.

Laser sources can be either butt coupled or lens coupled to fibers. Butt-coupling, even with index-matching, results in coupling efficiencies of about 5 percent for single-mode fibers and 20 percent for multimode fibers. Recall that LED butt-coupling efficiency is only a few percent. Since ILDs have small emitting areas and somewhat confined radiation patterns, lenses can be used to improve coupling efficiencies. By magnifying the emitting area, the radiating solid angle at the fiber core is made smaller, thus increasing the coupled energy. Two types of lenses were shown in Figure 4-5 to improve the coupling of LEDs to large-core multimode fibers. Figure 4-11 shows laser coupling to SM fibers using a self-focusing lens. The self-focusing lens is a single section of graded-index silica. If made a quarter wavelength long, it can be used either to collimate light from a point source or focus light. Alignment becomes more critical when microlenses like this are used. With a fiber lens (see Figure 4-5A), the need to carefully align a lens between the fiber and source is avoided.

All lenses can potentially reflect energy back into the laser cavity, which can cause performance degradations. Tilting components between

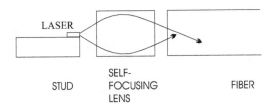

LASER

STUD | SELF-FOCUSING LENS | FIBER

Figure 4-11 *ILD coupling.*

the source and fiber sometimes is used to control these reflections. Any energy reflected from the tilted component does not go back into the laser's cavity. Source-fiber connections using this approach will sacrifice coupling efficiency for reflection reduction.

4.4.6 ILD Packaging and Reliability

A schematic of a packaged laser is shown in Figure 4-12. This ILD has a fairly simple package: fiber lens, fiber pigtail with connector, thermoelectric cooler, and back-end facet monitor. The fiber would be epoxied after alignment and the interior hermetically sealed. Electronic control circuitry is shown internal to the package. This would probably include signal drive circuits. A complete transmitter package would include drive, cooler and bias control circuitry. If this were a higher quality, 1300 to 1550 nm SLM laser transmitter package, the cost would be about $1000 US.

Package reliability depends on the quality of manufacture. A manufacturer will publish the results of life cycle tests and perhaps environmental tests. ILDs have their own set of reliability problems, which are often not under the direct control of the packaged transmitter manufacturer. The causes of degraded ILD performance and failure can be divided into three

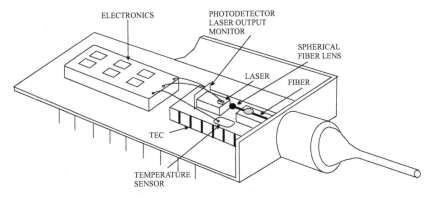

Figure 4-12 *Laser package.*

main categories: internal laser damage, ohmic contact degradation, and facet damage. Internal damage arises from the migration of crystal defects into the active region. Internal crystal defects grow into what are called dark spots/lines that interfere with stimulated emission. The quality of the chip manufacturing process dictates whether internal defects become important. Ohmic contact degradation increases the thermal resistance of the electrical connections, thus generating heat in the device. Excessive heat is to be avoided whenever possible. The quality of laser assembly dictates whether this becomes a problem. Facet damage can be caused by high optical power levels or long-term facet erosion. In both cases, the reflectivity of the device's ends (mirrors) diminishes, leading to performance degradations.

Higher ambient temperatures greatly reduce device lifetimes. Aging is a more serious problem in ILDs because they generally operate with high continuous current densities. This results in higher ambient temperatures if not compensated. Internal and facet problems in a laser are greatly accentuated when temperatures are elevated. Below a device ambient of 120 degrees centigrade, CW output power will remain constant for many years. Life testing using elevated temperatures was discussed in Section 4.3.6 for LEDs, and the approaches are the same for lasers. A unique indicator of eventual ILD life is the growth of the threshold current with elevated temperatures.

Voltage transients from static discharges are also troublesome for ILDs. Procedures to guard against ESD should be followed when handling ILDs, packaged or unpackaged.

Simple heat sinks are generally satisfactory for LEDs, but lasers require more precise and predictable temperature reduction and control. The thermoelectric cooler (TEC) included in Figure 4-12 consists of an array of n- and p- type semiconductor materials connected electrically in series and physically in parallel. Current through the TEC results in a temperature differential across the device and the ability to reject excess heat from a mounted component. The cooler can be designed to maintain a constant temperature or be placed in a closed loop under the control of the sampled laser output power (Figure 4-12). If the power changes, the assumption is that the device temperature has changed and the TEC bias is changed to compensate. An alternative to the TEC is to use a control loop on the current drive to achieve temperature compensation. The sampled laser output at the back facet would be used, through feedback, to change the drive current. This would act to keep the output power constant as the laser aged or the ambient temperature increased.

4.4.7 Vertical Cavity Surface Emitting Lasers (VCSEL)

A new type of semiconductor laser became available in the early 1990s, the vertical cavity surface emitting laser (VCSEL). Most of the lasers now used for high-speed local area networks (LAN) are 850 nm VCSELs. New 10 Gb/s Ethernet systems will use VCSELs and wavelength division multiplexing (WDM) in the 850 nm window. VCSELs offer a number of advantages over edge-emitting lasers: low cost, circular radiation pattern, low threshold current, higher efficiency, longer lifetime, and single-longitudinal mode operation. Unfortunately, the technology has yet to be transferred to the long wavelength windows. Edge-emitters still are used exclusively in the 1300–1600nm range, and also where high power is required.

A generalized cross section of an 850 nm VCSEL is shown in Figure 4-13. VCSELs in the 850 nm range are made from alloys of GaAs and GaAlAs. Recall that the heterostructure of an edge emitting ILD is deposited so that the emission occurs longitudinally. The heterostructure of the VCSEL is deposited so that the emission occurs vertically. Both types are fabricated on wafers using well established, semiconductor-manufacturing technology. Edge emitting ILDs have to be cleaved individually from the wafer and packaged before testing. VCSELs are tested while still in wafer form which leads to significant cost savings. With the VCSELs side by side on a wafer, a number can be purposefully fabricated as an array. Laser arrays are used for high power and optical cross-connect applications.

The active region of the VCSEL is only about 1–1.5 wavelengths long. In the 750–900nm wavelength range of current commercial VCSELs, the active region is only about a micron in depth. By comparison, the active region length of a Fabry-Perot edge-emitter is a few hundred microns. The short resonance cavity length results in single-longitudinal mode operation under most operating conditions. From equation 4-8, when the cavity length L is small, the spacing between modes is large. The required reflective mirrors on the top and bottom of the active region's cavity are distributed Bragg gratings. The bottom grating reflects 100% of incident light

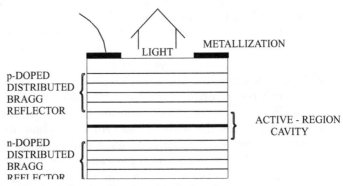

Figure 4-13 *Vertical Cavity Surface Emitting Laser.*

back into the cavity. The top grating is almost as efficient, allowing only about 1% of the light to escape as useful output power. The gratings have to be very efficient because the active region is small, which limits the amount of gain available to sustain oscillation. Bragg gratings act by reflecting energy at a wavelength determined by the spacing between the alternating layers of high-low refractive index material (Chapter 8). These are made of GaAlAs and GaAs

Much research is being expended to make VCSELs available at long wavelengths. The primary substrate material at these wavelengths, indium phosphide (InP), does not lend itself well to making Bragg reflectors. Lattice-matching is proving to be troublesome.

4.5   Optical Source Coherence

The output of a microwave oscillator in *continuous wave (CW)* mode is a sinusoid with a very small amount of noise. The noise causes a slight amount of spectral spread. This signal has a high level of time (temporal) coherence because its RF spectrum is nearly a single frequency. If the signal is radiated, the far field will have spatial coherence because it will exhibit uniform phase across the wave front. Finally, if the polarization is maintained throughout the propagation length the signal will also have polarization coherence. These are the three types of coherence an EM wave can exhibit: temporal, spatial, and polarization. Sources used by optical fiber systems are not yet as coherent as RF sources. However coherent, or semi-coherent, optical waves are needed to realize a few of the components used in optical fiber systems, e.g., interferometers, modulators, filters, lens couplers, and heterodyne receivers.

In a dictionary, the term *coherence* means to stick together. Whatever is coherent is "marked by an orderly or logical relation of parts that allow recognition or comprehension". A coherent EM wave is stable and predictable when displaced in time or space. Coherence is very high in signals from electronic oscillators. Of the two semiconductor optical sources, LED outputs are not coherent while lasers offer a limited degree of coherence. Gas and solid-state lasers can be very coherent, but they are not well suited to optical fiber systems.

Temporal coherence can be visualized by considering the phase relationship of a sinusoidal wave at two separate locations on the propagation path. As time progresses, the wave is coherent if the phase difference stays constant between the two locations. A completely incoherent wave would have no phase correlation, even over a very short period of time, or a short distance. A completely coherent sinusoidal wave would have zero noise modulation. A completely incoherent wave would be nothing but noise, because there would be no correlation from one instant to the next. The output of lasers can be temporally coherent, but an LED output will always be incoherent. Recall that lasers have narrow spectral linewidths and LEDs have very broad linewidths. This points out the inverse relationship between spectral linewidth, $\Delta\lambda$, and temporal coherence which is given by:

$$\tau_c = \text{coherence time} \approx \frac{1}{(2\pi\Delta f)} = \frac{\lambda^2}{(2\pi c \Delta\lambda)} \qquad 4.9$$

where $\dfrac{\Delta\lambda}{\lambda} = \dfrac{\Delta f}{f}$.

Using the velocity of propagation, the temporal coherence can be changed into length coherence. This is generally the more useful parameter because of the need to design components or systems that require EM wave coherence. The propagation distance over which the temporal coherence is maintained is determined by:

$$L_c = \text{coherence length} = (\text{velocity} \times \text{time}) = \left(\frac{c}{n}\right)\tau_c \qquad 4.10$$

Table 4-2 lists examples of coherence times and lengths for EM sources.

With the radio example, the significance of the long coherence time/distance is that, at a receiver, the incoming wave will be stable with a "predictable" phase. In a heterodyne receiver, the type normally used in radios, a coherent local oscillator frequency is mixed with the incoming signal. There will be no phase variation at the detector over the short period of time needed for accurate demodulation.

Some of the examples in Table 4-2 help illustrate the significance of coherence, or lack of it, in optical fiber systems. The single longitudinal mode, DFB laser is typical of those used in current, dense wavelength division multiplexed (DWDM) applications. The frequency spacing between channels in DWDM currently can be as low as 25 GHz. The 1 GHz linewidth is required if the channel's signal is to remain inside the assigned band. The laser source also has to be very stable. Couplers and filters used to join and separate these wavelengths must have small packages. The term *footprint* is often used in reference to package size. Inside the packages, the optical integrated circuits are very small. These types of components often rely on interference phenomena in their operation. Obtaining useful interference patterns, with good resolution, requires optical sources with sufficiently long coherence lengths. For example, one type of optical modulator separates an incoming laser signal into two paths. One path imparts a 180° phase shift if a "0" is to be sent or no phase shift if a "1" is to be sent. The two signals are then recombined. If they are in phase ("1"), there is an optical output from the modulator. If out of phase ("0"), there is no output. A

Table 4-2 *Source Coherence Examples—Temporal.*

SOURCE	$\Delta\lambda$ or Δf (Freq. or Wavelength)	COHERENCE TIME	COHERENCE LENGTH
Radio Oscillator	1 kHz	$0.16(10^{-3})$s	air: 48 km
He-Ne single-mode gas laser,600nm	0.64 MHz or 0.0076 nm	$2.5(10^{-6})$s	fiber: 500m
He-Ne multimode gas laser,600nm	0.16 GHz or 0.00038 nm	10^{-9} s	fiber: 0.2m
DFB semi. laser, 1500nm external mod.	1 GHz or 0.0075 nm	$0.16(10^{-9})$s	fiber: 3.2cm
GaAs multi-0.24(10^{-3})m mode, 850nm semi. laser	138 GHz or 1 nm	$0.12(10^{-11})$s	fiber:
LED, 850nm	12,200 GHz or 30 nm	$0.13(10^{-13})$s	fiber: $2.6(10^{-6})$m

coherence length of at least a centimeter would be needed to be able to construct this type of device. Consider also trying to use this laser to detect a 2.5 Gb/s pulse in a heterodyne receiver. The pulse period is $0.4(10^{-9})$s. The laser's coherence time is $0.16(10^{-9})$s, which is shorter than the pulse length. The phasing difference between the local oscillator and the incoming signal would not be stable, and the pulse might not be detectable.

Spatial coherence refers to phase consistency across a wavefront. The uniform plain wave presented in Chapter 2 was spatially coherent. The E-field had the same amplitude and phase throughout a plane perpendicular to the direction of propagation. Spatial coherence dictates the degree to which an interference pattern will be generated when two portions of the same wave are combined. In the modulator described above, both temporal and spatial coherence are required to obtain the interferometer operation. The EM waves within the two fibers were single mode and spatially coherent. This is why multimode fibers are not used in this type of component.

In another example, a uniform plain wave can be focused to a small spot using a lens. As an aside, diffraction effects caused by the lens will make the spot slightly larger than ray theory would predict. An LED cannot be similarly focused because it is not spatially coherent. A source can have temporal coherence but not spatial coherence. The radiation from a laser that has multiple transverse modes but only a single longitudinal mode will have temporal coherence but be spatially incoherent.

Polarization coherence refers to the source's ability to maintain a given state of polarization. The light from lasers is polarized because the cavity reflections are polarization sensitive. Coherent receivers require that the incoming signal have a fixed state of polarization. Optical receivers currently use only photodetectors, which are sensitive to intensity or incoming power, not polarization. Also, most of the optical fibers in use will not maintain a launched state of polarization. As the complexity and bit-rate of fiber systems increases, however, the state of polarization throughout the fiber link becomes more important (Chapter 8). Some component insertion losses are slightly dependent on polarization.

4.6 Erbium-doped Optical Amplifiers (EDFA)

Signal amplification has long been available in electronic and RF circuitry. Unfortunately, all through the first decade of optical fiber use there were no optical amplifiers commercially available. Signals had to go through an O/E conversion, followed by electronic regeneration and an E/O conversion. The process was costly. An example system that relied on regenerators was the first submarine optical fiber cable system, trans-Atlantic-8 (TAT-8). It went into service in 1988 and was retired about 15 years later.

TAT-8 carried 296 Mb/s, NRZ, on each of two fibers in both directions. It stretched 5600 km from the US to Europe. Depressed cladding, single-mode fiber was used with 1310 nm multi-longitudinal mode laser transmitters. Electronic regenerators were spaced about 40 km apart. Transmitters, receivers and regenerators were powered over the cable. Because of their bulk and complexity these O/E and E/O conversions required large and costly repeater housings. Redundant components were also necessary to meet an outage requirement of onl;y three failures in a 25- year lifespan. A failure meant grappling for the cable, which is expensive and time consuming.

TAT-8 would have been a simpler and much less expensive system if optical amplifiers were available to take the place of regenerators. The erbium-doped optical fiber amplifier (EDFA) became available in the early 1990's and is now used extensively. Somewhat ironically, EDFAs are spaced about 40 km in submarine systems, about the same distance as the earlier regenerators. The big difference is that fibers with amplifiers carry bit-rates in the Gb/s range on each of many wavelengths. The protective housings are still referred to as repeaters, even though their function now is only amplification.

This raises the need to distinguish between the three basic types of signal-enhancing components: an amplifier, a repeater and a regenerator. An *amplifier* simply increases the voltage, current or power level of a signal. The EDFA is an optical power amplifier used on single-mode fiber systems. Any noise accompanying the signal is amplified along with the signal. An EDFA can amplify many signals in a wide band, which is one of its main advantages. An amplifier will also add noise and, possibly, distortion to the signal. A *repeater* performs an O/E conversion, but uses the resulting electrical signal to drive a transmitter. A repeater is a receiver and transmitter, back-to-back. Noise accompanying the signal is passed on and even added to by the repeater. *Regenerators* actually make a new signal that is a clean replica of the initially transmitted signal. A term, the *3-R's*, is used to describe the actions of a regenerator. The regenerator re-amplifies, reshapes, and retimes. The three functions are realized with electronics, but research is being pursued to achieve regeneration optically. Regeneration is required after a certain number of amplified spans. Too many cascaded amplifiers will result in a significant noise and distortion accumulation. The distance between regenerators is referred to as the *reach*. The reach in submarine cable systems is in the thousands of kilometers, while that in the terrestrial long-haul network is about 600 km. This difference basically is the reason that terrestrial systems have amplifiers spaced at about 100 km.

EDFA amplifiers can be used at three locations in an optical fiber system: at the transmitter, at the receiver, or in the middle of the line. A transmitter amplifier is used to boost an output, possibly for distribution to multiple locations. Also, a laser/modulator's output power might be too low and need amplification. A receiver pre-amplifier boosts the weakened signal at the end of the fiber, thereby increasing receiver sensitivity. With the growing applications for networking at the optical level many components are being added to the signal path. These all contribute loss, and the transmitter and receiver amplifiers are needed to compensate.

An EDFA line amplifier compensates for losses in long fiber connections. The gain of the amplifier basically equals the loss of a section of fiber plus added optical components. The amplifier does add noise to the signal, called amplified spontaneous emission noise (ASE). The noise will accumulate with each amplifier section, which then requires regeneration of the electrical signal after hundreds of kilometers. ASE noise is about 30 dB down from the output signal in a single amplifier. The gain of an EDFA is about 30 dB. It varies with the input, and drops as the input rises. This effect is called saturation, and is present in all amplifiers. Output powers can be as high as a few hundred milliwatts. The spectral width is broad. Initially EDFAs were developed to cover the C-band, 1530 to 1565 nm (Chapter 8). Now, EDFAs are available also for the L-band, 1565 to 1625 nm.

A basic EDFA two-stage line amplifier is shown in block diagram form in Figure 4-14. The amplification process takes place in the two coils of silica fiber. Their cores are doped with the element Erbium. The fibers are only a few meters in length. Recall that the semiconductor laser (ILD) used injected current to excite electrons to higher energy states. This energy was then transferred to the desired, oscillating signal. EDFAs use injected light energy at specific wavelengths, 980 and 1480 nm. This causes the erbium atoms in the fiber cores to jump to excited states. The incoming signal stimulates the atoms to drop to a ground state, in the process achieving an energy transfer. The pump energy is essentially transferred to the 1550 nm signal. Since the pump wavelengths are shorter, they have higher energy than a signal photon.

In the example amplifier of Figure 4-14 the first stage fiber is forward pumped at 980 nm and the second stage fiber is reverse pumped at 1480 nm. There are some performance advantages to using this particular combination. Pumping forward at 980 nm produces lower ASE noise. Pumping in the counter-direction achieves higher gains. Amplification takes place over a broad range of wavelengths, but it is not equal over the whole band. This is the reason for the gain compensation function between the two gain sections.

In an EDFA operated at relatively low output power, the gain will peak at about 1535 nm and drop off with wavelength. The difference over the C-band can be as great as 10 dB per amplifier. For a wavelength division multiplexed system (WDM), that uses a number of channels spread over the C-band, the differential from the top to the bottom channel could be significant. The gain compensation stage attempts to equalize these gain differences. Fortunately, the gain difference diminishes as the amplifier's power output increases. WDM systems require higher total output powers so that each channel can have power sufficient to realize a satisfactory signal/noise

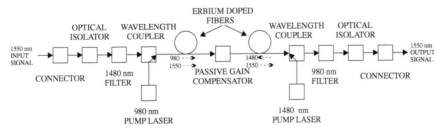

Figure 4-14 *Erbium Doped Fiber Amplifier.*

ratio. The gain compensation stage is set at manufacture. Adjustable or automatic compensation capability is still being developed. Optical networking, discussed later, requires this type of dynamic compensation because wavelengths will be dropped added or routes will be reconfigured.

The connectors, isolators, filters, wavelength couplers, and gain compensators are discussed in Chapter 6. Two other basic approaches to obtaining optical amplification are available but currently account for only a small percentage of the market. These are semiconductor optical amplifiers (SOA), and Raman amplifiers (Chapter 8).

Optical Detectors and Receivers

5.1 Introduction

A signal undergoes an E/O transformation at a fiber optic transmitter and a corresponding O/E transformation at the receiver. This chapter discusses the O/E process, with emphasis on the first two blocks in that flow: the semiconductor optical detector, or photodiode, and preamplifier. The two together are referred to as the receiver's front-end. With the growing need for commercial systems operating at 10 and even 40 Gb/s, the design of front-ends has become more important and technically challenging.

A complete optical receiver is more complicated than an optical transmitter. The signal from the fiber will be low in power and distorted because of the source and fiber characteristics. The electrical circuitry following the O/E transformation attempts to overcome some of these problems. In Chapter 4, the electrical circuits used to drive the optical source were not discussed. It was assumed that preceding the LED or ILD were drive circuits of sufficient bandwidth and output. Their design and realization is not overly troublesome because signal levels are high, and there are few noise or distortion problems.

In contrast, the circuitry that the semiconductor detector feeds directly into is very important in determining overall system performance. These two components, detector and preamplifier, are the dominant source of noise added to the signal in the receiver. As a result, the design of the front-end is critical in determining receiver performance. Front-end sensitivity is defined as the minimum detectable optical power in dBm. The sensitivity is determined by the required system signal-to-noise (S/N) ratio. The optimum choice of the type of photodiode and its preamp is dictated by the system requirements and constraints. These include, but are not limited to: wavelength, fiber length and type, bit rate or type of analog signal, noise objectives, cost objectives, acceptable circuit complexity, reliability, and environment. For example, a front-end chosen for a *local area network* (LAN) would not be satisfactory for an underwater cable system.

Optical detection techniques are presented in Section 5.2. Photon detectors are best for optical fiber systems because of the need for sensitivity. Section 5.3 discusses the construction of the two main types of photodiodes in current use, PINs and APDs. Section 5.4 compares PIN and APD performance. Section 5.5 discusses three basic preamp designs, low impedance, high impedance and transimpedance. Commercially available front-ends use combinations of the two diodes and three preamp designs.

5.2   Optical Detection

Optical energy can be detected by thermal and photon detector devices. The former detect changes in temperature as light is absorbed. The temperature dependency is transformed to an electrical signal for processing. These are useful in optical fiber sensor systems. Optical fiber communication systems use photon detectors. Photon detectors transform photons into electrons (in photoemissive tubes) or electron-hole (E-H) pairs (in photoconductive semiconductors). Both of these drive a current through an external circuit for subsequent applications.

Photoemissive devices are photon detectors that generate a single electron when a photon hits a metallic surface. When a photon with sufficiently high energy strikes, an electron will be dislodged and collected by a higher potential anode. It then becomes part of a current flow. Astronomers, for example, use high-sensitivity photoemissive vacuum tubes. A photon hitting the surface of the tube's cathode can dislodge an electron. The cathode material has a "work function" that the photon's energy has to meet or exceed in order for an electron to be dislodged and become part of an external current. Multiple anodes (called *dynodes*) are used in photomultiplier tubes to increase the number of electrons. This same effect is achieved in a semiconductor device called an *avalanche photodiode* (APD). Instead of dynodes, the APD uses a small layered section of an extremely high electric field to generate new E-H pairs.

Photoconductive devices are photon detectors that respond to light with an increase in conductivity. As the conductivity increases, the electrical resistance decreases and current through the device from an external power supply increases. There is much interest in adapting this type of detector to new fiber systems.

Photodiodes, are preferred for photon detectors in current optical fiber communication systems. Photodiodes are particularly well suited to the requirements of fiber systems because they are

- low in cost,
- small in size,
- constructed of materials compatible with opto-electronic circuit integration,
- sufficiently sensitive to light,
- relatively low in additive noise,
- an efficient transducer of O/E,
- fast in response time.

Photodiodes are reverse biased p-n junctions, made of various light sensitive semiconductor materials. The basic p-n junction is used in photodetectors in the following ways: as an abrupt junction, with a large intermediate intrinsic region (PIN diode), with large terminal voltage and large intrinsic region (APD) and as part of a transistor amplifier (phototransistor). The PIN and APD are preferred for fiber systems.

A digital optical receiver block diagram is shown in Figure 5-1. This chapter concentrates on the first two blocks, the optical diode detector and its preamplifier. The front-end and linear channel sections are similar for both analog and digital receivers. Both signals require a front-end that maximizes sensitivity and bandwidth. The differences between analog and digital receivers will be discussed in more detail in Chapter 7. Notice that the optical detector diode is reverse biased, and a diode designed as an optical source is forward biased.

The detector diode generates a photocurrent that is directly proportional to the impinging optical power. When the incoming optical power is modulated, the detector will follow these variations, translating them into changes in the output current. The diode becomes a current source. As mentioned in earlier chapters, the optical source is assumed to be intensity modulated. Other modulation approaches will require more complex receivers (Chapter 8).

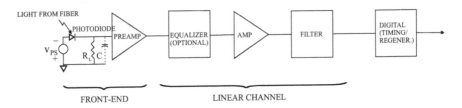

Figure 5-1 *Optical receiver block diagram.*

There are four main front-end performance parameters to consider: sensitivity, bandwidth, signal/noise ratio (S/N), detector signal output level. Sensitivity is determined by the detector characteristics and the required S/N. The load on the photodiode primarily determines the latter three. The value of the load resistor (R_L) is chosen in combination with the preamp's input resistance. As with any RF receiver, the pre-amplifier must add very little noise to the signal. The RC load on the detector is a combination of the controlled value of parallel resistance and parallel capacitances from the diode and preamp input. The equalizer can be used to compensate for this (RC) first-order rolloff in the frequency response if it is excessive. Equalization is generally needed for a high impedance preamp but not for a low impedance preamp (Section 5.5). The filter serves to maximize the S/N ratio by rejecting high-frequency additive noise. New generation systems also have equalizers and error detection/correction in the receiver to compensate for fiber induced degradations (Chapter 9).

Optical fiber photodiodes in current use are fabricated from silicon, germanium, GaAs, InGaAs, and InGaAsP. The choice depends on the wavelengths to be detected. A photon will penetrate these semiconductors and, if it has sufficient energy, generate an electron-hole (E-H) pair. These materials have different band-gap energy levels and are optimally responsive at different wavelengths (Chapter 4). Silicon and GaAs find use in the lower window (800 to 900 nm) and InGaAs and InGaAsP in the long wavelength window (1300 to 1600 nm). Germanium can be used across both bands, but is not preferred because of lower quantum efficiency and higher noise. The tertiary and quaternary alloys differ from those used for optical sources in their molecular ratios and fabrication. In current system designs, the silicon PIN and APD are preferred at short wavelengths and InGaAs/InGaAsP PINs at long wavelengths.

Transmitters and receivers usually are paired, since their designs need to complement each other. A manufacturer will generally market a transmitter/receiver together as a *transceiver*. This approach is logical and cost-effective since a transmitter and receiver are required at both ends of a link and need to be compatible. Front-ends are assembled as individual components, hybrid circuits on a substrate, or as ASICs. The form is primarily dictated by the bandwidth requirements. There is significant interest in developing and deploying 10 Gb/s and 40 Gb/s optical fiber systems. These require opto-electronic integrated RF circuits to meet the demanding bandwidth requirements (Chapter 8).

5.3 Photodiode Construction

Although an abrupt p-n junction can be made into an O/E converter, doing so is not efficient. A p-n junction detector is illustrated in Figure 5-2.

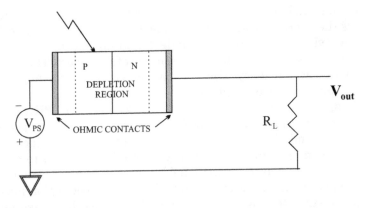

Figure 5-2 *PN junction photodiode.*

Conversion of photon energy to conducting electron energy takes place primarily in the intermediate depletion region. The depletion region is formed initially when holes and electrons from the two regions combine at the junction. Excess holes left behind form a positive space charge layer at the junction on the n side. Excess electrons left behind form a negative space charge layer on the p side. A potential barrier results between the p and n doped regions across the depletion region. This barrier forms the familiar 0.7-volt internal potential that must be overcome when a forward biased silicon diode is used in an electronic circuit.

Now assume no external field is applied. In Figure 5-1, $V_{PS} = 0$ and the circuit is open. Assume a photon penetrates the depletion region, generating an E-H pair. The electron is drawn to the p-layer of the n region, and the hole is drawn to the n-layer of the p region. They move at a drift velocity. The potential difference is enhanced. This is the *photovoltaic* effect used in solar cells. In the characteristic curves for a photodiode (Figure 5-6), this region falls in the quadrant where the diode current is negative while the naturally generated terminal voltage is positive. According to traditional circuit analysis, this means that a stand-alone cell is an energy source and could deliver power to a load. A typical silicon solar cell will have about 0.7 volts at its terminals and be able to yield a current of about 0.8 mA per square cm of surface when brightly illuminated.

In the p-n photodiode of Figure 5-2, the depletion region is small. The number of transformations from photons to E-H pairs is limited. Light is absorbed within a few microns of penetration depth. Each of the materials used for photodiodes has a characteristic response called the *absorption depth,* which is a function of wavelength. O/E conversion is increased if this depth is increased. Carriers can be generated in the regions outside the depletion region, but they result in a slower moving diffusion current. This increases the average transit time for the device, thus narrowing the bandwidth. In trying to penetrate to the depletion layer, a photon can also be prematurely absorbed without generating useful carriers. This results in a device with low efficiency. The O/E capability is greatly improved if an extended junction region is provided.

This approach yields the *positive-intrinsic-negative* (PIN) diode illustrated by the schematic in Figure 5-3. Here the depletion region becomes much larger, and the absorption depth in the device is increased. The intrinsic region is usually very lightly doped. If connected to an external load and a reverse polarity electric field is applied, the generated E-H pairs will provide the desired external current. The intrinsic region must have a uniformly distributed high electric field. This means that when fabricated, there must be a protective initial layer that the incident photons strike and penetrate. An important design objective is to make this layer thin or unresponsive to the wavelength's photons to keep the early absorption of photons low, thus adding to the quantum efficiency.

A long wavelength, mesa-construction PIN is illustrated in Figure 5-3. This type of photodiode is easily manufactured and inexpensive. All photodiodes are fabricated in layers. In this diode, the InP layers absorb only a small amount of radiation even though they are relatively thick. Note that a heterostructure is used, but not to provide signal channeling as in optical sources. Recall that a heterojunction occurs when two dissimilar III-V alloys are joined that have approximately the same lattice constant but dif-

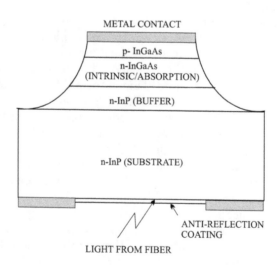

Figure 5-3 *Mesa structure PIN diode.*

fering band gaps. InGaAs provides efficient absorption and is preferred over Ge in the long wavelength window. InGaAsP is sometimes used in the initial layer because it leads to lower InGaAs dark currents.

The depletion region thickness is a few 10s of microns thick. The depletion region will extend slightly beyond the intrinsic region. This diode would have circular symmetry to match the fiber's cone-like radiation pattern. The fiber is butt coupled to the larger photodiode surface and almost all the available fiber output light collected. The term *photon bucket* is a commonly used descriptor for this application. From a practical viewpoint, detectors are made only slightly larger in the surface area than the fiber. The area directly affects the capacitance of the diode, which in turn, negatively impacts the device's bandwidth. A gap will exist between the fiber and the detector's surface. An anti-reflection coating is applied on the surface of all photodiodes to improve efficiency. The coating is a quarter of a wavelength thick at the diode's design wavelength, which reduces the Fresnel loss from 33 percent to about 5 percent.

The construction of a short wavelength silicon *avalanche photodiode* (APD) is schematically illustrated in Figure 5-4. Key differences between the APD and PIN are the high terminal voltage and the extra layering. The intrinsic section still contains most of the active depletion region. APD bias voltages are in the 150 to 300-volt range for Si, 20 to 40 volts for Ge, and 50 to 100 volts for InGaAsP. The avalanche, or multiplication region is the narrow added p-layer. Dark current (noise) is kept at reasonable levels in these diodes by separating the multiplication and absorption regions. The high

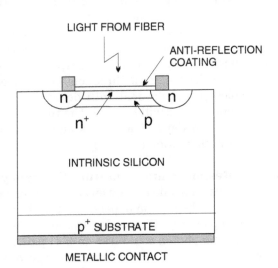

Figure 5-4 *Planar silicon APD.*

electric field accelerates the E-H pairs initially generated by a photon. Collisions occur with other electrons and holes, thereby giving multiplication. The effect is called *impact ionization*. The increased current is generated prior to encountering the load resistor, which gives the APD a sensitivity advantage over the PIN (Figure 5-7). Since the load resistor is a primary noise source in detectors, the increased photodiode current in the APD leads to improved S/N. This current amplification factor is labeled M in the equations below. The multiplication factor (M) is dependent on the diode's reverse breakdown voltage and the bias. Using multiplication for the signal also multiplies the (shot) noise contributed by the APD. Sensitivity or S/N is improved with M only up to the point where this shot noise equals the added preamp shot noise. The optimum multiplication factor is usually in the range of 10 to 100, depending on diode material.

Photon detectors in the near-infrared where optical fiber systems operate have significant additive shot noise. At wavelengths beyond a few microns, photon detectors are often cooled to reduce this noise. Cooling is not desirable for fiber system receivers because of the added cost and complexity. The gain of APD receivers is temperature sensitive. To keep APD gain at its optimum, the bias voltage is usually adjusted through a control loop that responds to temperature drift.

5.4   Photodiode Performance

The following three categories are used to describe the performance of photodiodes: spectral response and quantum efficiency, sensitivity and additive noise, and frequency and pulse response. To determine overall performance of the front-end, recall the four main performance parameters: sensitivity, bandwidth, S/N, and signal output. These parameters are addressed for photodiodes using the three combined categories. For S/N and signal output, which primarily deal with detector/preamp dependencies, simplified equivalent circuits will be used. The three basic preamp designs, low and high input impedance transistors and transimpedance, are discussed in some detail in Section 5.5.

5.4.1   Spectral Response and Quantum Efficiency

An electron falling from one energy level to another could give up a photon. The wavelength of the photon depends on the change in energy, which is called the band gap energy (Chapter 4). In the reverse O/E direction, the energy of an impinging photon could free an E-H pair. Remember that this is always a statistically based transition so the operative word is "could." Since transitions occur randomly, there will be an unavoidable amount of noise accompanying these processes. With optical sources, this showed up as spectral smear. With photodetectors it is quantum noise. Quantum noise is the lowest possible level of noise attainable in a photon detector.

The E-H pairs drift to the positive/negative reverse biased terminals and become part of the generated current. The energy in the photon is

$$E_P = hf = \frac{hc}{\lambda} \qquad 5.1$$

where h = Planck's constant = 6.63×10^{-34} joule-sec, f = optical frequency. This incoming energy must be greater than the band gap. Since the photon's energy is wavelength dependent, there will be a cutoff wavelength above which E-H pairs cannot be freed. This is given by

$$\lambda_{max} = \frac{hc}{E_G} \qquad 5.2$$

where E_G = band-gap energy in electron-volts (EV). Table 5-1 lists the approximate cutoff wavelengths for commonly used semiconductor photodiode materials. Silicon, with its 1.14 micron cutoff, is useable only in the short wavelength window from 800 to 900 nm. Germanium has a potentially broad range of wavelengths but is not as efficient an O/E transducer as silicon. Ge also generates significant internal shot noise. The InGaAs molecular composition listed is the preferred alloy for long wavelength PIN diodes. Its carriers have high mobility, which makes it attractive for high data rate, wide bandwidth optical photodiodes. Additionally, GaAs technology offers the advantages of low noise, wide bandwidth FETs and opto-integrated circuits.

InGaAsP is used with a number of different molecular weight combinations, depending on the desired wavelength range. The cutoff wavelength can be pushed out as far as 1.8 microns if needed. The response to incoming light for three common photodiode materials is plotted in Figure 5-5. Note how the response drops abruptly near the high-end cutoff wavelength. At

Table 5-1 *Photodiode cutoff wavelengths.*

MATERIAL	CUTOFF WAVELENGTH (μm)
Silicon (Si)	1.14
Germanium (Ge)	1.6
$In_{0.53}Ga_{0.47}As$	1.7
$In_{0.7}Ga_{0.3}As_{0.64}P_{0.36}$	1.4

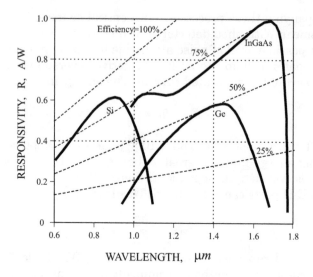

Figure 5-5 *Photodiode responsivity.*

low wavelengths, the absorption depth in all the materials decreases. This makes photons less likely to generate E-H pairs and the response again drops off, but more gradually.

Responsivity (R) is used as the measure of a detector's ability to turn photons into electrical current. Since a photodiode directly converts light to current, it is defined as

$$R = \frac{I_{out}}{P_{in}} \qquad 5.3$$

An APD will simply multiply the output current by a factor M, thus multiplying the responsivity by M. The quantum efficiency (η), plotted in Figure 5-5, is independent of the incident power at a given photon energy (wavelength). It is derived from

$$\eta = \frac{\text{Number of E - H pairs generated}}{\text{Number of incident photons}} = \frac{I_{out}/q}{P_{in}/hf} = \frac{Rhf}{q} \qquad 5.4$$

where I_{out} is in Coulombs/sec, q = electron charge = -1.6×10^{-19} Coulombs/electron, P_{in} is in Joules/sec, hf = photon energy in Joules/photon. Dimensionally, this yields electrons/photon. Note that the efficiency lines in Figure 5-5 are straight and rising. As the wavelength increases (frequency decreases), the photon energy decreases. As a result, the responsivity increases given the same ability in the material for a photon to generate an E-H pair. Using Equations 5.3 and 5.4,

$$R = \frac{\eta q}{hf} \qquad\qquad 5.5$$

The plots of responsivity in Figure 5-5 do not include the improvement available with anti-reflection coatings. For example, silicon will have its efficiency increased to 95 percent at 900 nm, giving a peak responsivity of about 0.63 A/W.

Typical photodiode I-V curves are plotted in Figure 5-6. In electronics, a diode's behavior in the upper right quadrant of the I-V characteristics is of most interest. The diode is forward biased and current flows as if the diode is a load, or passive device. With photodiodes, the lower left quadrant is of the most interest. The reverse bias will be in the range of 5 to 30 volts for a PIN diode and 20 to 400 volts APD. At a given wavelength, responsivity is constant as the input power changes. The resulting current change is then linearly related to any optical power input change (Equation 5.3). If the incoming light power increases by a factor of 2 (3 dB), the output current will increase by a factor of 2. This linear relationship is illustrated by the regular spacing of the flat current output lines in the negative quadrant. Photodiodes are linear over a wide range. The constant output current as a function of terminal voltage (at a fixed optical power input) indicates that the photodiode can be viewed as a constant current source with a very high output resistance.

A dc load line is superimposed on the example characteristic curves in Figure 5-6. A very simple equivalent circuit is represented: a supply voltage

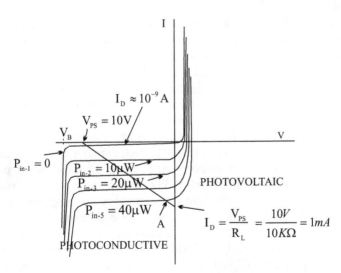

Figure 5-6 *Photodiode I-V characteristics.*

(V_{PS} in Figure 5-1) of 10 volts and load resistor (R_L in Figure 5-1) of 10k in series with an ideal photodiode. If an optical ac signal were received, the photodiode would respond along this line.

Note that at zero optical input power, the diode still outputs a *dark current*. Dark currents depend on the diode's construction and material. They are approximately: 1 nA in silicon, 10 nA in InGaAs/InGaAsP, >100 nA in Ge. In a digital front-end, a digital zero would mean zero incident light. Only the dark current would be present in the output, which would then constitute a noise current. In an analog front-end, a bias resistor would be used to place the quiescent operating point somewhere along the load line. Similar to a conventional electronic diode, when the reverse voltage becomes too high, the diode will suffer breakdown and possible damage. Also, if the input optical power is very high the output current saturates.

The load line offers insights into the resulting dynamic range of the front-end. This value is the difference between the maximum and minimum allowable optical power levels required for satisfactory system performance. The maximum allowable optical power is that signal level which results in the onset of receiver non-linear distortion and/or saturation. For the example in Figure 5-6, assume the optical power maximum occurs at point A (40 microwatts or –44 dBm). Assume further that the system requires an average power of 2 microwatts or –57 dBm to meet the required S/N. The dynamic range then becomes 57 – 44 = 13 dB.

5.4.2 Sensitivity and Additive Noise

Front-end *sensitivity* was defined as the minimum detectable optical power, established in value by the required S/N. The lower the absolute value of sensitivity, the longer can be the connecting link from transmitter to receiver. The signal level at the detector input is primarily a function of the output power of the source and the attenuation of the fiber. Receiver sensitivity depends mainly on the amount of added noise from the front-end. The sensitivity of PIN front-end designs is generally limited by the choice of preamp. An APD has higher sensitivity because of the internal signal gain. The drawback is that it requires more complex and expensive circuitry.

In the ideal case of no detector dark current, shot noise, resistor thermal noise, and amplifier shot noise, the detection sensitivity is only limited by the statistical nature of the conversion process. The statistical process describing this performance is called a *Poisson distribution*. The resulting sensitivity under these ideal conditions is referred to as the *quantum limit,* which is a baseline upon which front-ends are compared. The quantum limit represents the best sensitivity attainable for a given S/N objective. In this ideal case assuming equally probable "1"s and "0"s, to detect an optical pulse with a bit error rate (BER) of 10^{-9}, an average of 10 photons per bit is required. This applies to all photon detectors and is the quantum limit for the given error rate. The required average received optical power for this BER of 10^{-9} is

$$P_{QL} = 10hfB \qquad\qquad 5.6$$

where B = bit rate. Figure 5-7 compares the quantum limit to the expected performance ranges for PIN and APD front-ends at 1550 nm for the conditions of average received optical power and 100% detector efficiency. The vertical axis gives the required incident optical power and sensitivity. No distinction is made relative to the type of diode material, preamp, or wavelength. Note that the best front-ends, from a sensitivity viewpoint only, are about 15 dB above the quantum limit for a BER of 10^{-9}. Long wavelength APDs do not perform as well as short wavelength silicon APDs because of high dark current and excess noise. As benchmarks, silicon APDs at 0.85 microns operate at about 15 dB above the quantum limit, and PINFETs about 25 dB above. Germanium APDs at 1.3 microns operate about 20 dB above the limit. The other long wavelength devices fall in between.

A receiver's sensitivity value depends on the required system S/N. The *signal/noise ratio* (S/N) gives a relative measure, rather than an absolute measure, of a system or sub-system's noise performance. The sensitivity can be found by starting with the S/N objective and working back from the point in the receiver where all noise sources can be calculated and added. The S/N power ratio is given by:

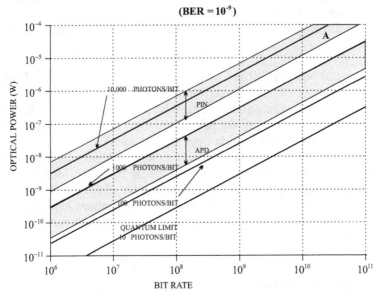

Figure 5-7 *Minimum optical signal power (1550 nm, average power, 100% efficiency).*

Figure 5-7 illustrates a fundamental problem in designing very high bit rate receivers. Currently 10 Gb/s systems are becoming operational, and widespread deployment of 40 Gb/s systems is on the horizon. Based on current data, point A in Figure 5-7 represents the performance of a 40 Gb/s, PIN diode receiver. This system would have a sensitivity of only about –12 dBm in achieving a 10^{-9} BER performance. Assuming a laser source with + 3dBm coupled power to a fiber, only 15 dB is available for fiber attenuation and component losses between the transmitter and receiver. If a fiber/component average loss of 0.5 dB/km is assumed, the allowable span length is then 30 km. One way to improve on this situation is to use an optical EDFA preamp (Section 4-6). A signal level of +6 dBm would be typical at the detector input after optical preamplification. At this level of input, significant saturation will be avoided and the amplifier plus thermal noises will be essentially equal. Assuming 30 dB gain for the EDFA, the optical signal power into the optical preamp would be –24 dBm, a 12 dB improvement over the non-amplified approach. At 0.5 dB/km this would allow another 24 km of span length. One drawback, besides added cost and complexity, is the added amplifier noise. This will slightly degrade the receiver's overall S/N.

$$\frac{S}{N} = \frac{\text{Signal power caused by photocurrent}}{\text{Device noise power + thermal noise power}} \qquad 5.7$$

The receiver's intended use in a system determines the required S/N. For example, a digital receiver might require a S/N of 15 dB, while an analog video receiver might require 40 dB. The signal power in the numerator is the average of the output current squared times the load resistor in Figure 5-1.

$$\text{Signal Power} = P_S = \langle i_{out}^2(t) \rangle R_L = I_{out}^2 R_L \qquad 5.8$$

The signal power is a time average value, which is easy to obtain since the signal is deterministic (describable by mathematical expression). The noise powers are mean square values because the noise is statistical in nature. Amplifiers following the preamp are not assumed to add significant noise, because the signal level is high at that point.

Receiver noise is made up of two basic components: noise generated by electron motion in devices and thermal noise from the load resistor. The photodiode's noise power will be a summation of the dark current noise mentioned previously and shot noise caused by random electron motion in the diode. The latter is dependent on the diode's average current. The noise current squared is given as

$$\langle i_{ns}^2(t) \rangle = 2q(I_{out} + I_D)B \qquad 5.9$$

where B = noise bandwidth (generally the receiver bandwidth), q = Electron charge = -1.6×10^{-19} C, I_D = dark current. The load resistor contributes thermal or Johnson noise current squared of

$$\langle i_R^2(t) \rangle = \frac{4kTB}{R_L} \qquad \qquad 5.10$$

where k = Boltzman's constant = 1.38 x 10^{-23}, T = noise temperature (Kelvin). Multiplying all of these by the load resistor value gives the noise power, and the resulting S/N becomes

$$\frac{S}{N} = \frac{I_{out}^2 R_L}{2q(I_{out} + I_D) BR_L + 4kTB} \qquad \qquad 5.11$$

An additional noise contribution from the preamp will be added in Section 5.5. If the load resistor is made large, the thermal noise contribution is lessened on a relative basis. A higher value of load resistor would also produce a higher output voltage to drive the preamp. In the other direction, a smaller resistor is needed if the dynamic range is to be maximized. Note that the slope of the load line in Figure 5-6 would be increased and the vertical intercept made greater with a lower value resistor. The increased slope increases the maximum allowable received optical energy. Also, the available bandwidth will be increased if the resistor is decreased in value (Section 5.4.3).

With an APD, the signal power will be increased by M^2, the square of the current multiplication factor. The shot noise power from the diode is increased by a factor M^{2+x}. The parameter M^x is called the *excess noise factor*. The value of x is a function of the APD material, which is approximately 0.4 for silicon, 1.0 for Germanium and 0.65 for InGaAs. The APD equivalent equation to Equation 5.11 then becomes:

$$\frac{S}{N} = \frac{I_{out}^2 M^2 R_L}{2q(I_{out} + I_D)BM^{2+x} R_L + 4kTB} \qquad \qquad 5.12$$

Again, the preamp noise will be added in Section 5.5.

If the realized S/N falls below the required value, because of an unanticipated additive noise, the system's signal may be degraded but still useable. Assuming the photodiode, preamp and signal bandwidth are not changed, an increased S/N objective would call for a higher absolute value of receiver sensitivity. This assumes also that the shot noise is not dependent on the signal level. In all front-ends, shot noise is dependent on the signal level (Equation 5.9), however, which serves to illustrate one of the many trade-offs faced in front-end design.

When used as an objective, the S/N states the maximum acceptable additive noise disruption that a specific baseband signal can tolerate. Distortions to the signal would have a separate, independent requirement. With the digital receiver in Figure 5-1, for example, sensitivity would generally be quoted as the optical power needed to achieve a 10^{-9} BER (for a given photodiode-preamp combination) in the absence of other degradations.

When a specific system S/N requirement is not specified, the S/N is customarily taken as 1.

5.4.3 Frequency and Pulse Response

The front-end's frequency and response time performance is determined by the RC time constant and/or carrier transit time. The first generates a 20 dB/decade rolloff, or single-pole frequency response. Using the 3 dB frequency, this rolloff can be translated to a pulse exponential rise and decay in the time domain (Chapters 3 and 4).

The other parameter is the average transit time required for electrons, once generated, to exit the photodiode and become part of an external current. E-H pairs generated in the depletion region move at a drift velocity to their respective terminals. Carriers generated in the surrounding layers move at a slower rate, the *diffusion velocity,* towards the intrinsic region where they become part of the faster drift current. Together they dictate the diode's transit time. The simplified schematic in Figure 5-1 does not include all circuit elements needed to analyze the RC response of front-ends. A more detailed and accurate equivalent circuit is shown in Figure 5-8. The model combines the diode's internal electrical equivalent circuit with the load resistor and the input impedance of the preamp. The diode series resistance (R_S) can be ignored because it is very small. The diode's large shunt resistance (R_{DP}) is the reason the diode is a current source in the equivalent circuit. At a fixed value of optical power, the current is constant over a wide range of negative bias voltage. The diode's internal capacitance is small (0.2 to 1 pF). The capacitance is very important to the front-end performance, and is a focus of attention when fabricating diodes. The diode's capacitance is directly dependent on the surface area of the diode, and inversely on the width of the depletion layer. This is why the diode surface is designed to be only slightly larger than the fiber, and one of the advantages to the wide depletion region used in both the PIN and APD.

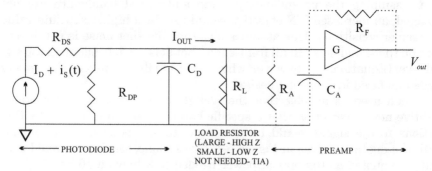

Figure 5-8 *Front-end equivalent circuit.*

A diode will have a natural 3 dB frequency $(R_{DP} | | C_D)$ that is generally very high. The parallel load resistor needed to generate drive for the pre-amp unfortunately lowers this frequency considerably. The resistor will have a value ranging from <100 ohms to > 1 megohm, depending on the chosen preamp circuit and its input impedance. The choice of preamp type is an important consideration, and involves several tradeoffs (Section 5.5).

The parallel resistors and capacitors comprise a first-order circuit for which the 3 dB frequency is given by:

$$f_{3-dB} = \frac{1}{2\pi R_T C_T} \qquad 5.13$$

where $R_T = R_{DP}$, R_L, R_A in parallel, $C_T = C_D$, C_A in parallel.

The load resistor in Equation 5.10 now has become part of R_T. The feedback resistor in Figure 5-1 is used only for the *transimpedance design* (TIA) discussed in Section 5.5.

As an example of a typical frequency response, assume the capacitances are each 1 pF, the load resistor is 10 kilohms and the amplifier's input resistance is 1 megohm. The resulting bandwidth is 1.6 GHz. A pulse response's rise time from the circuit parameters alone becomes (Equation 3.4).

$$t_r = \frac{0.35}{f_{3dB}} = \frac{0.35}{1.6 \times 10^9} = 0.22 \times 10^{-9} \text{ sec} \qquad 5.14$$

This is the 10 to 90 percent rise and fall time on a detected pulse due to the (first order) RC circuit. If the load resistor increases, the circuit response time increases, which increases pulse distortion. The equalizing filter in Figure 5-1 is used to compensate for front-end bandwidth limitations caused by high values of load resistance or capacitance.

A diode's total pulse response can also be dependent on internal carrier transit times. The E-H pairs generated in the extremes of the intrinsic region take a finite time to cross. Also, carriers generated in the p and n regions surrounding the intrinsic region diffuse even more slowly to the depletion region before they drift across. In silica, the depletion region speeds are 8.4×10^6 cm/s for electrons and 4.4×10^6 cm/s for holes. Diffusion velocity is about 100 times slower. The p and n doped regions are always made narrow to increase quantum efficiency and reduce average transit times. Transit dependent response times are typically less than 1 ns in silicon photodiodes. Comparing this to Equation 5-14 shows that the circuit and transit time responses might be equally important in some front-ends. Transit times for APDs are somewhat slower. In receivers intended for very high data rates, the circuit response will probably be under control leaving the (internal) diode transit time to limit bandwidth. If one component is at least one-fourth of the other, their effects should be combined. Since they are independent variables, they can be added on a root-sum-square basis.

For example, an InGaAs PIN photodiode with depletion width 1.5 microns will have a transit-time limited bandwidth of approximately 20 GHz. Increasing the depletion width to 5 microns to increase efficiency will reduce the transit-time dependent bandwidth by half. In the opposite direction, the increased width will decrease the diode's capacitance, thus potentially increasing circuit bandwidth. This is another example of the trade-offs presented in designing front-ends.

5.4.4 Photodiode Summary

Table 5-2 summarizes the characteristics of the primary photodiodes used for optical fiber systems. The values listed are representative, approximate and typical. Individual devices will deviate from these numbers – sometimes significantly. Practical devices have wide ranges for many of these characteristics. The bandwidths of diodes intended for very high data rate systems are much greater. Equivalent rise/fall times are listed because first order circuit models have been used throughout our analyses as a basis for relating time response to bandwidth limitations. The bias voltages for PINs range from 5 to 30 volts.

5.5 Preamplifiers

There are three broad preamp categories: *high impedance amplifier, low impedance* transmission line/amplifier and *transimpedance amplifier* (TIA). Each provides a different load to the photodiode. Each amplifier has unique advantages and applications. The transimpedance amplifier has become the preferred approach for long wavelength high data rate systems.

Table 5-2 *Photodiode characteristics.*

	Units	Silicon (Si)	Germanium (Ge)	Indium Gallium Arsenide (InGaAs)
PIN				
Peak Response Wavelength	μm	0.9	1.4	1.6
Peak Responsivity (R)	A/W	0.6	0.5	1
Dark Current	nA	1	100	10
Rise/Fall Time	ns	0.75	0.2	0.1
Bandwidth (0.35/(Rise/Fall time))	GHz	0.5	1.75	3.5
APD				
Peak Response Wavelength	μm	0.9	1.4	1.6
Avalanche Gain	------	100	20	30
Peak Responsivity (MxR)	A/W	60	10	30
Dark Current	nA	0.5	400	150
Rise/Fall Time	ns	1	2	0.25
Bandwidth (0.35/(Rise/Fall time))	GHz	0.35	0.175	1.4
Bias Voltage	V	275	30	25

A high impedance amplifier provides a high detector signal level to the preamp and also reduces the relative contribution of load resistor thermal noise. The load resistor is 100 K to a few megohms. This combination maximizes the S/N (Equations 5.10 and 5.11). The disadvantage is that the resulting low (RC) 3 dB rolloff frequency limits the bandwidth (Equation 5.12) and must be equalized (see Figure 5-1). This leads to an enhancement of the low frequency energy and potential saturation of the receiver, thus limiting the dynamic range. The preamp can be designed with a wide dynamic range to help compensate. The detector characteristics will change over time, which means that the equalizer will need adjustment.

Silicon bipolar transistors and MOSFETS and GaAsFETs are used as preamps in high impedance and TIA front-ends. The FETs provide superior noise performance over BJTs. Silicon PINFET integrated circuit (IC) front-ends are widely used for the lower wavelength window. A silicon PINFET has bandwidth sufficient for moderately high bit rates (500Mb/s). The availability of GaAs microwave FETs has allowed the realization of very wideband, high impedance IC preamps for the long wavelength windows.

A low impedance preamp can be used to obtain a wide frequency band while avoiding the need for equalization. As a rule of thumb, the resistor choice in determining the need for equalization is guided by

$$R_T < \frac{1}{2\pi C_T B} \qquad 5.15$$

Equation 5.14 helps explain the need for integrated circuit front-ends to achieve very high bit rate receivers. Assume a hybrid circuit front-end with the photodiode connected by short coax to the preamp. A parallel 50-ohm load resistor and 50-ohm coax connected to a 50-ohm preamp would require a $C_T \approx 2$ pf. As an approximation, this assumes the bit rate and bandwidth are both equal to B $\approx 3 \times 10^9$ Hz. This is a reasonable value of total capacitance. The bandwidths (hence bit rates) are limited in discrete or hybrid front-ends. The 50-ohm load resistor at the detector would be needed to avoid reflection problems and allow reasonable component spacing. This low impedance approach would result in higher relative resistor thermal noise and a smaller signal level, thus a worsened S/N. Low impedance preamps are used mainly with APDs.

The most effective preamp for high data rate systems is the transimpedance amplifier (TIA). In Figure 5-8, this configuration is realized by using an op-amp instead of a conventional high or low impedance amplifier. The need for equalization is avoided, as is any dynamic range limitation. The load resistor (R_L) is dropped and R_A is sufficiently high so that it can be ignored. This type of circuit is a current to voltage converter. The gain is given in volts/amp or ohms. With a simple ideal op-amp calculation, it can be shown that the preamp output voltage depends on the product of the diode output current and the feedback resistor. The feedback resistor (R_F) becomes the load resistor. A wide bandwidth is obtained if both the op-amp gain and the feedback resistor (R_F), are large. Because the op-amp input voltage is very low, the entire supply voltage appears across the diode, regardless of optical power input. The load line in Figure 5-6 thus becomes vertical. TIAs have slightly poorer sensitivity than high-impedance front-ends.

All preamps will add noise to a receiver. For simplification, it can be assumed that the added noise is similar to thermal noise. This allows the use of a noise figure for the preamp. An amplifier's noise figure, or noise factor, is defined as

$$F = \text{Noise Figure} = \frac{S_{input}/N_{input}}{S_{output}/N_{output}} \qquad 5.16$$

If the amplifier were to add no noise, the noise figure would equal 1 or 0 dB. The S/N equations, Equations 5.11 and 5.12, now become respectively:

$$\frac{S}{N} = \frac{I^2_{out} R_T}{2q(I_{out} + I_D)BR_T + 4kTBF} \qquad 5.17$$

$$\frac{S}{N} = \frac{I^2_{out} M^2 R_T}{2q(I_{out} + I_D)M^{2+x} BR_T + 4kTBF} \qquad 5.18$$

These equations are simplifications but can be used to make approximate performance calculations. The use of a noise figure for the preamp actually approximates and combines two independent transistor noise sources: a series voltage source and a parallel current source. FETs and BJTs differ in the values and relative roles of these sources. Further, the equivalent transistor voltage noise source is a function of C^2_T which is another reason to minimize the diode's capacitance.

Even though there are many choices for a front-end, two directions have emerged through experience. If the 800 to 900 nm window is going to be used at relatively low data rates, the silicon PINFET or APD is preferred. Since the low wavelength window has emerged in the marketplace as the home for lower speed, relatively short systems, the silicon PINFET is the winner of the two. The APD will give about 10 dB more sensitivity, but costs more.

In the long wavelength window, the marketplace calls for higher speed, relatively long systems. InGaAs/InGaAsP PINs with TIA are generally preferred. An APD will give a few more decibels of sensitivity at a significantly higher cost. These front-ends will be discussed further in Chapter 7 in the context of total system performance.

6

Optical Components

6.1 Introduction

Optical fiber use in telecommunications has matured beyond a basic point-to-point system. The optical signals by themselves are now multiplexed with other signals, switched, and re-routed. Achieving this requires many different optical components. The major components discussed here are:

- connectors
- splices
- couplers/splitters
- wavelength division multiplexers (WDM) and filters
- circulators
- isolators
- attenuators
- switches
- modulators

The first four of these devices are passive. A *passive device* does not require energy from a power supply or an added signal input. All devices have fiber input/output ports and are fabricated from fibers, integrated optics, or micro-optic components in various combinations. The devices that use fibers alone are the cheapest to produce. Sometimes these components require externally applied fields. Integrated optic devices use optical waveguides implanted in substrates. Microoptic devices use lenses, filters, and gratings. Because microoptic components must be small and stable, they are the most costly to produce. However, they also can be made to be more wavelength sensitive, which makes them attractive for optical multiplexing/demultiplexing.

Connectors afford flexibility in routing fibers and are generally used in terminal equipment. They are attached to the fibers either in the field or the factory. *Splices* are used to make permanent connections between fibers and can also be field or factory installed. Splices are used in place of connectors whenever possible because connectors have more insertion loss, and are more prone to failure. *Couplers,* which are generally bi-directional, are used to combine and split signals. *Wavelength division multiplexers* and demultiplexers are similar to couplers/splitters except for their wavelength sensitivity. They pass or reject optical bands of wavelengths, often down to widths as narrow as 0.2 nm (25 GHz at 1550 nm). *Filters* are made either of deposited layers or reflection gratings. *Circulators* are three-port devices that pass signals in one direction from one port to the next. *Isolators* are one-way optical devices used primarily with single-longitudinal-mode lasers. They prevent optical reflections from entering back into the laser cavity. *Attenuators* are used to decrease high-level signals, and also to facilitate testing. *Optical switches* are used to route the optical signal. Currently, optical switches do not operate fast enough to allow bit-by-bit control. This is the reason there are still no optical computers. *Modulators* are used most often with SLM semiconductor lasers so that the signal does not have to be injected into the laser cavity. This avoids some of the problems with very high bit rate systems that use semiconductor laser sources.

6.2   Connectors

Connectors and splices both function to connect two fibers with a minimum of loss. They differ in that the connector is demountable and that the splice is permanent. Attenuation is the only forward direction transmission problem caused by connectors and splices. There are no wavelength dependencies that would lead to signal distortions. Reflected energy in the reverse direction can cause problems with high bit rate systems, however. The fiber-to-fiber transmission losses discussed in the following sections appear in both connectors and splices, but with different severities. Splices can be made with better fiber-to-fiber alignment accuracy, which means less loss. The same loss factors developed for silica fibers apply to plastic fibers. The losses will be different because plastic fibers are always step-index and have larger NAs.

The basic factors used to quantify all connection losses are divided into external factors and internal factors. External factors are under the control of the connector's designer/manufacturer. The higher the quality of design/manufacture, the lower the loss from these factors. Internal factors are controlled by the properties of the fibers being connected. If they come from different reels or manufacturers, they are likely to have slightly dissimilar profiles, numerical apertures, or geometry. Multimode fiber connections are largely affected by internal factors and single-mode fiber connections by external factors. This is logical since single-mode connections require submicron accuracies in aligning the two fibers.

These factors apply to both multimode and single-mode connections and are listed in Table 6-1. Graded index multimode fibers (GMM) with parabolic cores have been assumed in this discussion. Their use now seems to be universally preferred over step-index multimode fibers for communication systems. Step-index single-mode fibers, with Gaussian or bell-shaped fundamental mode patterns, have also been assumed. Single-mode fibers have their energy concentrated along the axis of the core, while MM fibers have energy spread out over both the core and cladding. Single-mode fiber connection sensitivity to the factors in Table 6-1 is markedly different than that of GMM fibers. The use of index matching fluid to reduce Fresnel reflection loss in connectors is not assumed in deriving the losses for these factors.

Cross sections of the two basic types of connectors used commercially are sketched in Figure 6-1. All connectors are intended to align the fibers as accurately as possible on a repeatable basis. The butt-coupled connector (Figure 6-1A) is easiest to attach in the field and is the type most widely used. The precision parts are the ferrule and the sleeve. The other basic type of connector uses lenses between the fibers. The expanded beam approach (see Figure 6-1B) uses grin-rod lenses to collimate and then focus the light beam. Sensitivity to lateral offset of the two fibers is diminished with the expanded-beam connector but precise control of the fiber's position relative to the lens is required. Angular misalignment can also be more troublesome because the focused output can be displaced from the fiber

Table 6-1 *Connection insertion loss factors (connectors and splices).*

EXTERNAL:
Lateral offset
End-gap
Angular tilt
Surface quality
Reflection
INTERNAL
Profile mismatch (MM)
Fundamental mode mismatch (SM)
Core radii & NA mismatch (MM)

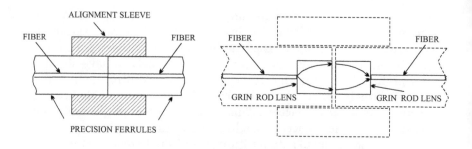

ALIGNMENT SLEEVE
FIBER
FIBER
PRECISION FERRULES
FIBER
FIBER
GRIN ROD LENS
GRIN ROD LENS
FIG 6-1A BUTT COUPLED w/FERRULE
FIG 6-1B EXPANDED BEAM

Figure 6-1 *Basic connectors.*

core. This connector's main advantage is that it allows easier connection to other devices such as splitters, multiplexers or filters, because the beam is collimated in the middle. This type of connector also is less sensitive to large gaps between the fibers and contaminants, such as dust. The drawback with this connector is the precision required of the fiber-lens relationship and the resulting higher cost.

Any interdependence of the loss factors in Table 6-1 is generally ignored because the required analyses would be too complex. As a first approximation then, the losses can be summed in dB as long as each is relatively small (<1 dB). Many different types of connectors are available that satisfy a wide variety of applications. If they are of good quality and attached properly, total insertion loss will be less than half a dB.

6.2.1 *Multimode Connectors*

An accurate insertion loss value for a GMM connector is difficult to predict, mainly because the transmitted energy is not consistently distributed across the fiber's cross section. A multimode fiber's core will have a large diameter relative to the transmitted wavelength. The fiber will thus be able to sustain many modes, but not all of them for long distances. The higher order modes will have more energy on the periphery of the core. As a result, they become susceptible to higher losses at bends and discontinuities, such as connectors and splices. Also, energy will propagate for a short distance in cladding modes. These have a large amount of energy outside the core and are susceptible to loss. As a result, they will not propagate over long distances.

This leads to the use of two modal fill descriptors to define the effects of connectors/splices on multimode fibers, *full-fill* and *steady state*. In the full-fill condition, each of the modes that can be sustained (even for limited distance) has equal energy. Full-fill would occur just after energy has been coupled from a non-coherent source or fiber that has a broad radiation pattern, for example a surface emitting LED. The term *uniform fill* is also used for this condition. The attenuation of any connection's imperfection, for example core-to-core lateral offset, is maximized under full-fill conditions. More energy appears at the edges of the fiber core, which then is lost when traversing the connection. In the steady-state condition, the cladding and higher order mode energy is diminished or absent. The steady state condition occurs after transmission through a few meters of fiber.

Connectors are measured and specified using the steady-state condition. To achieve steady-state, full-fill launch is followed by a mode filter that strips the higher order modes. The fiber's modal structure then will be in a pseudo steady-state condition. Half of the connector is attached to the transmitting fiber. As a result, the energy launched into the receiving fiber's half connector will be in a confined cone. There will be no appreciable higher order mode losses in the receiving fiber because of the steady-state modal fill in the transmitting fiber. When power is measured after the connector, a true measurement of insertion loss is obtained.

In a system installation, however, this same connector might be attached to a very short LED pigtail, and the transmitting fiber would be over-moded. The connector would contribute an effective insertion loss higher than that measured under the steady-state measurement technique described above. Data on commercial multimode connectors should be read carefully to fully appreciate the published insertion loss values. The numerical results presented in this chapter describing multimode connector/splice losses are based on mathematical calculations that assume the more conservative full-fill condition. Analyses that lead to numerical results for these loss factors are more easily made assuming a full-fill.

Figure 6-2 sketches the fiber-to-fiber relationships for the first three external factors listed in Table 6-1: offset, end-gap, and tilt. A butt-coupled connector is sketched, but the same factors apply to an expanded beam

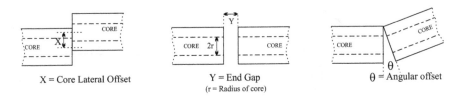

| X = Core Lateral Offset | Y = End Gap (r = Radius of core) | θ = Angular offset |

Figure 6-2 *External connection factors (all fibers).*

connector. With the expanded beam the sensitivities are lessened somewhat but the cost is increased.

The losses in butt-coupled connectors for these three external factors are presented graphically in Figure 6-3 for a parabolic GMM fiber. The parameters, x, y, θ, and r, are defined in Figure 6-2. The independent (horizontal) variables are ratios of these parameters so that a common scale can be used. Note that gap loss is always fairly low. Small gaps result in little loss of peripheral energy. Lateral or offset misalignment has the greatest loss potential. This is not overly troublesome since the cores of MM fibers are large and connectors are manufactured to close tolerances. To achieve an offset loss less than 0.5 dB, the cores have to be misaligned by less than 25 percent of the core diameter. Assuming GMM fibers with 100 micron core diameters, the cores can be offset by up to 25 microns. In Section 6.1.2, on single-mode connectors, this same type of requirement leads to the need for submicron machining, which can be costly. For angular misalignment, the variable is the fraction of the fiber's acceptance angle (θ_A). The acceptance angle is derived from $\sin \theta_A$ = numerical aperture = NA. In the fraction, NA(0) means the on-axis NA.

If an acceptance angle were measured under full-fill launch conditions, the NA would be high. After about 50 microns, it would lower to a value described as the *NA equivalent*. As an example, the full-fill NA might be about 0.35 but would lower to an NA equivalent of about 0.24. The NA stabilizes at the NA equivalent at longer lengths. This is the value of NA (and acceptance angle) generally used to evaluate coupling efficiencies.

The results in Figure 6-3 assume a full-fill condition. At a normalized value of 0.3, the steady-state condition would give results about 1 dB lower

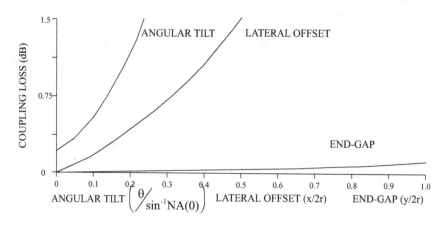

Figure 6-3 *GMM fiber external factor losses.*

for angular misalignment and about 0.3 dB lower for offset. Gap loss is not greatly affected by the fill condition because the gaps are relatively small.

The fourth external factor in Table 6-1 is the quality of the fiber surfaces. Light will be scattered and lost from a surface that has significant specular imperfections. Surface variations should be no greater than 0.025 microns. There are two ways to prepare a fiber for the connection process: cleaving or polishing. A good quality connector application procedure requires both. Splicing often requires only cleaving. The process of attaching connectors to fibers calls for the following sequence, which is abbreviated here for simplicity.

1. Remove fiber buffer, clean to expose cladding,
2. Epoxy fiber in ferrule, and
3. Cleave exposed fiber end and polish.

The last step is crucial in obtaining a high quality surface. As anyone who has cut a piece of glass knows, making a sharp surface mark can lead to a clean break. A clean break means a flat surface, generally orthogonal to the fiber's axis. The term *scribe and break* is applied to the cleaving activity. After the fiber is scored, an exact amount of tension is applied to obtain a clean break. The cleaving activity should leave no rough edges of significant size. Some connection surfaces are intentionally cleaved and/or polished at a slight angle to the normal. This approach reduces reflections. The polishing activity is referred to as *lap and polish*. The fiber is polished using successively finer grades of polishing paper or surfaces. Machines are used to polish multiple fibers simultaneously. The fiber surface can be shaped through polishing to be concave, flat or convex. A *physical contact* (PC) connector uses a convex surface to allow the two fibers to touch. This eliminates reflections, but requires precise gap control. If the ends of fibers are compressed in the connector, their surface refractive indices change and the reflection problem remains. It has also been shown that polishing can increase the refractive index at the fiber's surface. This can make the reflection problem slightly worse.

The last external factor in Table 6-1 is reflection. With an air gap between the two fibers and smooth orthogonal surfaces, a single Fresnel reflection will result in a 0.18 dB loss per surface, 0.36 dB total for the silica-air-silica gap (Chapter 2). The use of index matching fluid, slanted face, or PC connectors helps alleviate this problem. Matching fluid is difficult to use with connectors, however, because they must be easily taken apart and reassembled.

Note that an additional, resonance, effect is listed in Table 6-1. If the optical source is coherent or semi-coherent, a small gap between fibers will result in a standing wave. The resonance effect can contribute additional loss. Laser sources will be more susceptible to this problem than LEDs. A standing wave has maximums and minimums. Since the exact amount of end-gap is unpredictable, a conservative estimate of the additional loss is 0.5 dB for all connectors. These reflections in single-mode connectors can also cause wavelength instability in the laser source.

The internal factors listed in Table 6-1 account for dissimilar fibers. As with the external factors, the numerical results that follow assume a full-fill condition and GMM fibers with parabolic core profiles. In Equation 2.17, which describes the parabolic index GMM fiber, the term $(r/a)^2$ gave the parabolic shape because of the exponent of two. A different graded index would have a different power on r/a. A more general term describing index shapes uses $(r/a)^s$, where $s = 2$ for the parabolic index. If the exponent $s = 1$, the core profile approximates a triangle. If the exponent is a large number, the core profile is step-index. Figure 6-4 plots estimated connection loss as a function of the deviation from a power of two based on the equation:

$$L_{\text{GMM-INT}}(s) = -10 \log \left(\frac{s_R}{s_T} \right)\left(\frac{s_T + 2}{s_R + 2} \right), \text{ for } s_R < s_T \qquad 6.1$$

Both exponents are close to 2. These results assume that the receiving profile has the lower exponent. If the receiving profile has a shape wider than the transmitting, there is zero loss from index profile deviation. The results also assume the radii and NA of the two fibers are identical. A step-index fiber connection will not have this loss factor.

A mismatch in radii (a), assuming the profiles and numerical apertures (NA) match, is given by

$$L_{\text{GMM-INT}}(a) = -10\log \left(\frac{a_R}{a_T} \right)^2, \text{ for } a_R < a_T \qquad 6.2$$

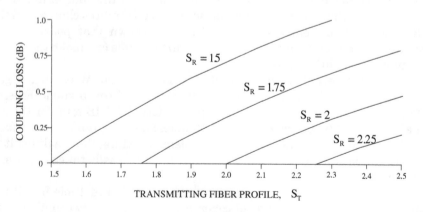

Figure 6-4 *GMM fiber internal profile losses.*

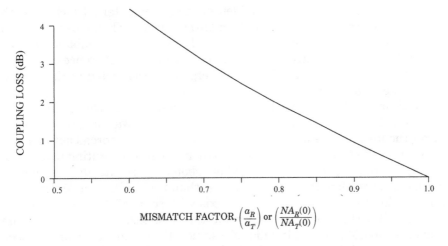

Figure 6-5 *GMM fiber core radii and NA mismatch losses.*

where a_R = radius of receiving fiber, a_T = radius of transmitting fiber. If the emitting fiber radius is smaller than the receiving fiber radius, there is zero loss. A mismatch in NA, assuming profiles and radii match, is given by

$$L_{\text{GMM-INT}}(\text{NA}) = -10\log\left(\frac{\text{NA}_R(0)}{\text{NA}_T(0)}\right)^2, \text{ for NA}_R(0) < \text{NA}_T(0) \qquad 6.3$$

where $\text{NA}_R(0)$ = axial numerical aperture of receiving fiber, $\text{NA}_T(0)$ = axial numerical aperture of transmitting fiber. If the emitting fiber NA is smaller than the receiving fiber NA, there is zero loss. With a step-index fiber, the axial NA is the core NA. Figure 6-5 shows how these squared relationships would translate into connection loss. Note that NA or radii departures of only 10 percent will give losses approaching 1 dB. These results are based simply on geometric relationships. Analysis of these mismatches using a Gaussian model for the energy distributions shows slightly lower values.

Comparing the external factor performances plotted in Figure 6-3 to those of the internal factors (Figures 6-4 and 6-5) leads to the conclusion that the latter will probably be more troublesome for GMM fiber connections. Typically, losses from internal factors alone will range up to 0.2 dB given a good quality connector.

6.2.2 Single-Mode Connectors

The performance of multimode connectors is not wavelength dependent. The results are dependent only on the modal fill, which was assumed to be full or uniform. Because of the large cores in MM fibers, the internal factors are potentially more troublesome than the external factors. With SM fibers external factors tend to be more significant. Their cores are much smaller, which makes misalignments relatively more troublesome. The internal factors (refractive index profile, radius, NA differences) are less troublesome because so much of the energy is concentrated in the central part of the core.

Multimode fibers have hundreds of sustainable modes because the cores are large relative to the propagating light's wavelength. For each mode, the maximum and minimum field intensities are spread out in a pattern across the face of the waveguide. When a fiber is operating with a single-mode, the wavelength and fiber geometry result in a value of normalized wavelength (V), which is less than 2.403 (Chapter 3). The maximum field intensity appears on the axis with some of the energy spreading into the cladding, depending on the spot size (w). For 1300 nm conventional SM fibers, it is valid to assume a Gaussian shape for the mode intensity. Recall that the spot size is the point on the Gaussian curve where the beam intensity (power) is $1/e^2 = 13.5$ percent of the peak central value. For dispersion shifted or dispersion flattened cores, the spot sizes are smaller because the core profiles are more complex (Chapter 2). The spot size of the fundamental mode increases approximately linearly with the wavelength for a given core radius.

Single-mode connection external losses can be derived from the following general equation:

$$L_{\text{SM-EXT}}(\text{dB}) = -10 \log K_1 \, (K_2 \backslash K_3) \, e^{-(K_4 K_5/K_3)} \qquad 6.4$$

where K_1 = reflection loss with air space

$$K_2 = \left(\frac{w_R}{w_T}\right)^2, \; w_R \text{ and } w_T \text{ are mode field radii}$$

$$K_3 = \left[\frac{1}{4}\right]\left(\left(\frac{y\lambda}{2\pi w_T^2}\right)^2 + (K_2 + 1)^2\right)$$

$$K_4 = \left(\frac{2\pi w_T}{\lambda}\right)^2$$

$$K_5 = (K_2^2 + 1)\left(\frac{x\lambda}{2\pi w_T^2}\right)^2 + 2K_2 xy \left(\frac{\lambda}{2\pi w_T^2}\right)^2 \sin\theta$$

$$+ K_2 \left(\left(\frac{y\lambda}{2\pi w_T^2}\right)^2 + K_2 + 1\right) \sin^2\theta$$

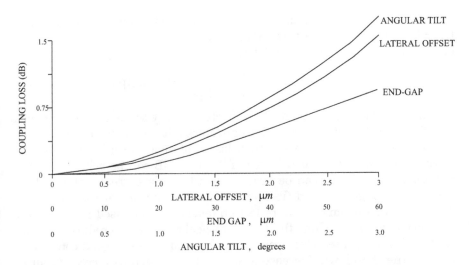

Figure 6-6 *Single mode fiber external factor losses (1300 nm standard sm fiber).*

Figure 6-6 plots loss estimates for the first three external factors, x, y, and θ shown in Table 6-2. Several assumptions were made in using Equation 6.4 to obtain the simplified results shown. First, the step-index, SM fiber characteristics, listed in Table 2-1 were used. This results in a transmitting mode field radius W_T of 4.98 microns at an operating wavelength of 1300 nm. The normalized frequency is 2.09, making the fiber operate slightly below the 2.403 higher order mode cutoff. The receiving and transmitting mode fields were set equal and the Fresnel reflection loss (K_1) set equal to 1.

Following the approach with the GMM fiber, the losses caused by each factor were calculated setting the other two to zero. For brevity, the equations that give these results will not be reproduced here. If the operating wavelength, core diameter, NA, or refractive indices were to change, these results would be slightly different. Lateral offset gives the greatest relative potential loss. For this reason, manufacturers try to keep core eccentricity to a minimum, thus helping to keep axes aligned.

Surfaces for SM connectors must meet the same 0.025 micron objective mentioned for MM fibers for surface scattering to be minimized. Reflections will cause the same Fresnel loss as MM fibers, i.e., 0.36 dB for the two surfaces. Single-mode connectors also have the additional reflection problem caused by gap resonance. The standing wave set up in a gap can, if conditions are proper, increase loss by as much as 0.5 dB. The effect is more likely to occur with SM fiber systems that use coherent lasers.

Semiconductor lasers are also sensitive to energy reflected back into their lasing cavity. When this energy is compared to the impinging signal energy, a *return loss* is derived from:

$$L_{RL} = -10\log\left(\frac{\text{reflected power}}{\text{incident power}}\right), \text{ in dB} \qquad 6.5$$

In Chapter 2, the ratio of the reflected energy to incident energy was identified as the reflectance (R). Connectors with air gaps and no correction for reflections will have return losses of about 14 dB $(-10\log(0.04))$. With PC connectors, return losses of 25 to 40 dB can be achieved. When index-matching fluids can be used, return losses are greater than 30 dB. Finally, a 5 degree sloped fiber end connector can achieve about 45 dB. The system requirements on reflection levels are discussed in Chapter 7. The use of index-matching fluid or a physical contact also reduces the connector's insertion loss while improving the return loss. Angled connectors, on the other hand, will increase insertion loss as well as increase connector cost.

Single-mode connector internal loss factors depend solely on the two fiber spot sizes. If Gaussian shapes are assumed for both, the loss K_2/K_3 in Equation 6.4 reduces to:

$$L_{SM\text{-}SPOT} = -10\log\left(\frac{2w_1w_2}{w_1^2 + w_2^2}\right)^2 \qquad 6.6$$

where w_1, w_2 = mode spot sizes of fibers. This assumes that all three of the external factors (lateral offset, end-gap, tilt) are zero. The loss formula is reciprocal, i.e., it does not matter which fiber has the larger spot size. Numerical results are plotted in Figure 6-7 and apply to all SM fibers where the Gaussian fundamental mode model is appropriate. Note that the second fiber has to have a spot size deviation of about 20 percent before the insertion loss rises to one-fourth dB. Single-mode connection losses due to internal factors alone will generally be < 0.1 dB.

Some of these loss factors, both external and internal, appear to offer a potential for large insertion losses. In the field, they do not occur simultaneously, however. Similar to GMM connectors, good quality SM connectors will have total losses < 0.5 dB.

6.3 Splices

All of the potential loss factors mentioned for connectors are possible with splices but are much better controlled. Insertion losses are significantly lower as a result. This is the reason splicing is generally preferred over connectorization. With the newest splicing equipment, losses are held

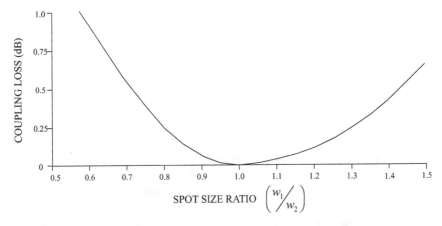

Figure 6-7 *Single mode spot size loss.*

routinely to < 0.1 dB for both MM and SM fibers. As a result, in assembling a link from transmitter to receiver, splices are chosen whenever possible in place of connectors. In a 100 km link, splices will be needed about every 2 km. Assuming splice losses of 0.05 dB/splice, the splices would add 2.5 dB loss to the required loss margin (Chapter 7). At high data rates this can be a significant issue because margins are small.

A successful splicing operation will achieve three basic goals: good fiber end preparation, accurate fiber alignment, permanent alignment retention. Fiber ends are prepared by cleaving a smooth surface on both fibers, perpendicular to their axes. Alignment to within very small tolerances is achieved through completely or partially automated techniques. Splices are permanently sealed and protected. The details depend on whether mechanical splicing or fusion splicing is used. Fusion splicing results in a continuous fiber connection and is used almost exclusively for SM fibers.

Mechanical splices use a number of approaches to ensure good alignment: V-grooves, tubes, sleeves, and rods. All of these have to be manufactured to precise dimensions to achieve good alignment results. The fibers to be spliced must have outer diameter and core dimensions that comply with the splice specifications. After the fibers are cleaned externally they are cleaved and sometimes polished. Alignment is achieved either passively, by relying on the precision of the splice component parts, or actively, using the techniques described below. Epoxies are used to provide both index matching and stability. Mechanical splices are more susceptible to degradations caused by the environment. Mechanical splices can also generate troublesome reflections because they are not always continuous. For high data rate direct detection systems, return losses > 30 dB are required, and mechanical splices are generally avoided.

In fusion splicing, fibers first have their protective coatings removed and ends scribed and broken. The fibers are mounted in the splicer and aligned. Alignment can be achieved passively, for example by using an eyepiece, or actively. Active alignment gives the best results. Two types of computer controlled, automatic alignment systems are available: *profile alignment* (PAL), and *light injection and detection* (LID), both of which yield excellent results. The PAL approach uses orthogonal mounted sensor systems that operate somewhat like an x-ray machine. The two cores are compared and their relative positions adjusted prior to fusion. The uniformity of each of the fibers' concentricity and ellipticity are important in achieving a low loss splice. The LID approach injects light just ahead of the intended splice and detects it on the other side. The fiber is bent in both locations to allow injection and detection. The fibers are positioned automatically through computer control by minimizing the measured loss. The fusion process for both methodologies takes place quickly in a controlled sequence. Either an electric arc or gas flame is used to melt the glass and bring about fusion. The fused splice will have about half the tensile strength of the silica fiber. Flame fusion gives a stronger splice and is preferred for submarine cable installations. The splices are recoated, reinforced and organized in a splice holder. One advantage of a fusion splice over a mechanical splice is its resulting small size, which makes it possible to combine many splices in a relatively small splice enclosure. Multiple splices can be accomplished simultaneously if the fibers are mounted at the factory in a spatially controlled array. Individual fiber PAL or LID alignment is not possible when splicing arrays. Contamination is a potential problem with all splicing operations. An air-conditioned, dehumidified environment is required. Often, contamination encroachment is unseen and unanticipated. Resulting transmission troubles may show up years later, without any apparent cause.

6.4   Couplers and Splitters

Couplers and splitters are used to subtract a portion or all of the mainstream signals for such uses as monitoring or rerouting. They can also be used for wavelength combining (Section 6.5). They are fabricated either by using fibers brought in very close proximity or with lens/mirrors in combination. Figure 6-8A gives a sketch of a fused biconical tapered coupler, and Figure 6-8B shows a Grin-rod lens coupler. The latter is an example of a *microoptics* device.

A coupler will be designed to give a fixed drop loss, P_A to P_D, which can range from 0 to 15 dB or more. If the drop loss were 0 dB, for example, the

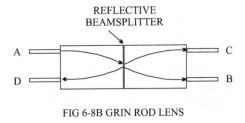

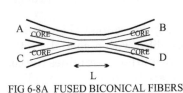

FIG 6-8A FUSED BICONICAL FIBERS

FIG 6-8B GRIN ROD LENS

Figure 6-8 *Directional couplers.*

signal output at the through port, B, would be zero and all energy would be transferred to the second output port. None of the input appears at port C except by reflection. This directional isolation leads to the descriptor *directional coupler*. Some small additional excess loss may occur as reflected energy at P_A and P_C. The design is valid over a limited wavelength range. Most couplers are bi-directional. As a result, the reverse direction would give the same performance.

Couplers can have more than one dropping port. The term *splitter* is sometimes used to describe this type of device. The devices in Figure 6-8 are 2×2 couplers. An $N \times M$ coupler would have N input ports and M output ports. This is often referred to as a star coupler. As an example, consider a 64×64 splitter (64 input and 64 output ports). If all ports were equally coupled, an input signal on one would appear down 18 dB (10 log 64) on each of the others. This type of device finds use in local area networks. Multimode couplers with many ports are made by fusing multiple fibers. Single-mode couplers with many ports are made using 1×2 couplers in cascade.

In the fiber coupler of Figure 6-8A, the wavelength and the coupled power level depend on the length of the tapered coupling region (L) and the degree to which the evanescent waves couple. Both parameters are fixed at the time of fabrication. The device is called a fused biconical tapered coupler because the input/output fiber radii are reduced and fused. The length of the coupling region is critical to achieving the intended performance. At a given wavelength, the degree of coupling is a sinusoidal function of the length of the coupling region. This wavelength dependency makes the coupler usable as a wavelength multiplexing device (Section 6.5). A gradual diminishing of the fiber's core radius reduces reflections. In multimode couplers, the input cores are gradually decreased in radius and the claddings are fused through the coupling region. Higher order modal energy travels in the common cladding of the coupling region. On the output side the cores diverge and increase back to their original radii. The higher order modal energy returns to the cores but now some energy has been transferred to the second guide, as desired. In SM fibers, light is coupled between the evanescent and overlapping parts of the fundamental modes. The core axial offsets core radii, and the length of the interaction region govern the degree of coupling.

The two Grin-rod lenses in the integrated optics coupler (Figure 6-8B) are each one fourth wavelength long and assembled with a partially reflective beam-splitter film in the middle. The beam-splitter determines the coupling ratio. Depending on the designed characteristics of the beam-splitter, some light is reflected back to port D, and the rest is transmitted to the through port B. Very little energy appears at A and C, as in the fiber directional coupler. Excess loss is about 1 dB for both couplers. The beam-splitter can be made wavelength sensitive which also makes this type of coupler capable of wavelength multiplexing.

6.5   Wavelength Division Multiplexers (WDM) and Filters

Spectral filtering is common in RF systems. In the future, optical channels on fibers will be multiplexed using filters and heterodyne techniques much like radio channels are frequency multiplexed today over the air. Optical systems using spectral filtering have not yet reached that capability, but they are rapidly improving. A great deal of R&D effort is currently being concentrated on using wavelength division multiplexing (WDM) to increase fiber capacity and improve optical networking flexibility. WDM is available for both single-mode and multimode fiber applications. Signals are sometimes transmitted in opposite directions (bi-directional WDM) on the same fiber using different wavelengths. Early WDM systems were designed to combine channels in the three main windows at 850, 1300, and 1550 nm. Single-mode fibers carrying four channels at the longer wavelengths are common in commercial use. Equipment for multiplexing 4, 8, and up to 160 channels is now available commercially. The ability to use hundreds of carriers on a single-mode fiber, greatly increases the capacity of fibers already in the ground. This section deals only with WDM devices and optical filters. System applications are discussed more completely in Chapters 7 and 8.

Data rates transmitted on a single optical carrier have grown in a very short period of time from 2.5 to 10 and now 40 Gb/s. The pulse streams carry time-division-multiplexed (TDM) signals, packet-switched internet traffic, stored data transfer, and fast Ethernet. The cost increases with the bit rate, but the cost/bit/km comes down as the bit rate increases for a system already in place. In the near future 40 Gb/s appears to present an electronic limit. Moreover, some studies have concluded that semiconductor lasers have a natural modulation frequency limit in the range of 50 GHz. Wavelength multiplexing allows continued growth because it combines many optical carriers on the same fiber. Also, the explosive growth in data traffic has made it economical and desirable to route high data rate signals through networks at the optical carrier level. Individual wavelengths, with their modulation intact, can be added, dropped, switched, or re-routed using WDM.

An international recommendation, ITU-T G.692, specifies *dense wavelength division multiplexing* (DWDM) channel multiplexing using spectrum slices of 0.4, 0.8, and 1.6 nm width (50, 100, and 200 GHz at 1550 nm). Equipment to space channels as close as 25 GH_2 is being developed. Using these guidelines, the C-Band from 1525 to 1565 nm could accommodate more than 100 high data rate channels, giving a total capacity >1Tb/s. Costs are an important issue for DWDM however. When channel spacing is required to be close, the laser transmitters and required precision optical components are more expensive. For very densely packed channels, the laser sources must be temperature controlled and externally modulated. Newly commercialized single-mode fiber, free of the water peak at 1385 nm, offers an inexpensive alternative to DWDM. Since more wavelength range would be available, the individual channels can be farther apart.

Coarse wavelength division multiplexing (CWDM) is being used to bring costs down while still offering significant increases in fiber capacity. The CWDM approach calls for 16 channels spaced at 20 nm through the entire 1300 to 1620 nm long wavelength region by using fiber free of the water peak. If conventional SM fiber with a water peak is used, the 1300 nm window could transmit 4 CWDM channels while the 1500 nm window could transmit 8 CWDM channels. The filters used in CWDM are less expensive than with DWDM. The central 3 dB pass band is a smaller percentage of the channel assignment, about 13nm out of 20. CWDM systems are also available on good quality multimode fibers. Eight 12 nm channels are used in the 850 nm window, each potentially transmitting 2.5 Gb/s. This multiplexing is used in local area network (LAN) and Ethernet applications.

The main concerns in using WDM devices are added device insertion loss and possible crosstalk. Insertion losses are of the order of 1 dB with current WDM devices, as long as they are not cascaded. Crosstalk shows up as an added interference when the desired signal is detected. Fortunately, intensity modulation/detection is rugged and does not require significant interference rejection: 25 to 35 dB is generally sufficient. All WDM devices are passive; most WDM devices function equally well as multiplexers or demultiplexers.

Similar to the design of electrical filters, optical filters can be bandpass, band-reject, high or low pass. Bandpass and band-reject filter designs are used almost exclusively for DWDM systems. They are manufactured using deposited thin films and/or gratings. The gratings can be either micro-optic, i.e. requiring lenses, or implanted in fibers. The filter's shape in the frequency/wavelength domain, which is essentially a transfer function, will have the usual sharp skirts and relatively flat passband. The target for unwanted adjacent and non-adjacent interference rejection is > 25 dB. The in-band amplitude characteristic should be flat to within a few tenths of a dB. A *figure-of-merit* (*FOM*) for DWDM filters is derived by dividing the bandwidth at the 25 dB rejection frequencies by the pass-band bandwidth at the 0.5 dB down frequencies. A DWDM filter designed for 100 GHz channels should have bandwidths of 1.2 nm and 0.4 nm respectively, giving a FOM of 3. A DWDM filter for 50 GHz channels should have bandwidths of 0.5 and 0.3 nm respectively, giving a FOM of 1.7. The lower the FOM, the more expensive the filter. Thin film filters are cost effective for 200 and 100 GHz channel spacing. Fiber Bragg grating filters are cost effective at 50 and 25 GHz channel spacing. Another consideration for filters is the dispersion characteristic within the filter's passband. Recall that the phase characteristic is not important with intensity modulation. A good quality filter will have < 50 ps/nm dispersion. From equation 3.2, assuming a laser with 0.01 nm linewidth, filter dispersion will not be troublesome.

The fused biconical fiber coupler in Figure 6-8A can be used as a single-mode WDM device to combine/separate two signals. One commercial device gives 25 dB crosstalk isolation and 0.25 dB insertion loss. A system that has eight CWDM channels spaced 20 nm apart between 1470 and 1610 nm could be built by cascading four of these couplers. This type of WDM device is inexpensive but does not offer the very narrow selectivity required for DWDM. Integrated or microoptic components have to be used to increase channel density further and reduce total insertion loss.

The coupler in Figure 6-8B uses a beam splitter that reflects some energy and passes the rest. Replacing the beam splitter with a layered filter yields a WDM device that can separate two independent wavelengths. Two wavelengths are input at port A. One will be reflected back to the C port and the other transmitted out the B port. Insertion losses are about 1 dB, and crosstalk < –25 dB. Layered film filters are made from thin films of material with differing refractive indices. If designed for more than one wavelength, the filter will have multiple layers. Depending on the width of the layer(s) and the wavelengths, reflections at the internal boundaries will enhance or cancel the impinging or transmitted signal(s). Thin-film filters have the advantage of being thermally stable.

The fiber coupler in Figure 6-8A can be realized using precisely deposited waveguides in a glass substrate. The same principles of operation apply;, i.e., the length of the coupling region and the degree of coupling determine performance. Figure 6-9A illustrates an integrated optics WDM device that uses a Lithium Niobate (LiNbO$_3$) substrate instead of glass. The index of refraction of Lithium Niobate can be changed with the application of an electric field. Light at one wavelength can be coupled completely to

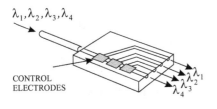
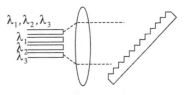

CONTROL
ELECTRODES

FIG 6-9A INTEGRATED OPTICS
(PLANAR LIGHTWAVE CIRCUIT -PLC)

FIG 6-9B REFLECTION GRATING

Figure 6-9 *WDM devices.*

one of the three parallel output ports through electrical control. The other wavelength(s) continue unhindered. This allows integrated optic wavelength selectors, switches, modulators and other devices.

Figure 6-9B illustrates another WDM device using micro-optics. All WDM filters depend on the principles of constructive and destructive interference. The resulting interference patterns that give the wavelength selectivity are generated either through the use of layered filters or reflective gratings. The reflection grating (also called a diffraction grating) in Figure 6-9B is a wavelength-sensitive mirror, formed by placing a periodic structure into a reflective material. The grating is at an angle relative to the collimated beam. The three different wavelengths are reflected at different angles. The lens focuses them in the focal plane at the different fiber locations. The wavelengths are combined and focused on A when radiating from ports B, C, and D. In the other direction, if all are radiating from A, for example, they are focused on their respective output fibers. Micro-optic devices can be fabricated to process a large number of channels for DWDM. These devices provide very good selectivity but are not cost-effective for a low number of channels.

Another filter that is finding important and increased application is the imbedded-fiber Bragg grating filter. A Bragg grating follows *Bragg's Law:*

$$\Lambda = \text{grating period} = m\lambda/2 \qquad 6.7$$

where λ = wavelength tuned by the grating, and m = order of grating. The grating period is the distance between the grating peaks. The order of the gratings used in imbedded fiber filters is generally 1. A higher order uses corrugations with wider spacing. The fiber Bragg grating is fabricated by irradiating the fiber's core with ultraviolet light that implants periodic "cells" of increased refractive index within the core. Signal energy transmitted through the fiber is reflected at the cell sites causing periodic constructive and destructive interference while the wave traverses the fiber filter. The wavelength dependency of these reflections gives rise to a resonance effect and tuning. Fiber Bragg gratings in use today are of this reflected wave type. The desired filtered signal is in the direction opposite to the incoming signal. Varying the periodicity of the index variations broadens the filter's bandwidth. This is referred to as *chirping*, but should not be confused with the chirping of a laser's output that sometimes occurs when it is pulsed on and off.

The distributed feedback injection laser (SLM-DFB, Chapter 4) uses a Bragg grating imbedded in the confinement layer to obtain a single longitudinal mode output. These gratings are generally second order because of problems in depositing a closely spaced first order grating in semiconductor alloys. A distributed Bragg reflector laser is also available (SLM-DBR), which uses externally coupled gratings in semiconductor material to obtain the required sharp tuning.

Fiber Bragg gratings can also be tuned. The spacing (Λ), and hence the resonance wavelength is sensitive to temperature and pressure changes. This results in some very important new applications: tunable filters and lasers (Chapter 8). Also, reflective fiber gratings are used to compensate for chromatic dispersion. By varying the spacing within the grating, the reflected wavelengths occur at slightly different times and can be used to compensate for the differential delays of the fiber.

An add/drop multiplexer/demultiplexer using fiber Bragg gratings and circulators is illustrated in Figure 6-10. Fiber Bragg gratings can also make use of the transmitted wave. Circulators are used to separate and combine the wavelength selected by the filter from the other wavelengths.

Multiple Fiber Bragg gratings are used in a DWDM device called an *arrayed waveguide grating* (AWG). Like micro-optic reflection gratings, they provide the advantage of separating channels in a single step with all

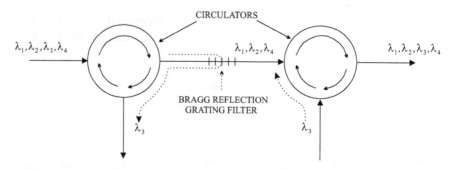

Figure 6-10 *Drop and mux/demux.*

outputs in parallel. This reduces the insertion loss problems caused by cascaded devices. An AWG can function as a multiplexer/demultiplexer, drop/add device, or wavelength router. Currently available AWGs can multiplex/demultiplex 40 and even 80 channels using a single integrated optics chip. Insertion losses for all channels range from 2 to 4 dB, and crosstalk is about 30 dB in an AWG.

An important factor in the growth of WDM use is the ability to amplify many optical signals simultaneously with the same device. The optical amplifier in widespread use is the *Erbium-doped-optical amplifier* (EDFA) used in the 1550 nm window. EDFA's can be used on SM fiber only. Their bandwidth can be made wide enough to handle a number of channels. A new optical amplifier, the Raman amplifier (Chapter 8), is now available which makes amplification of WDM channels at lower wavelengths also possible.

6.6   Circulators

The add/drop devices in Figure 6-10 use a circulator to direct wavelengths to and from the grating. Circulators are used in the same way at RF. Optical circulators use the Faraday rotation phenomenon discussed in Section 6.7 relative to isolators. Note that signal flow within the device is always in one direction. A circulator could be used with bi-directional transmission on a single fiber to separate the transmitted and received wavelengths. Insertion losses are about 1 dB and crosstalk isolation is better than 25 dB.

6.7   Isolators

Reflections are a serious concern at microwave and RF frequencies. If the transmission distance is a significant percentage of the transmitted carrier wavelength (sometimes taken as 10 percent), a reflection can cause troublesome echoes. This results in interference noise, standing waves, and power loss. Since optical wavelengths are much shorter, the problem of reflections will occur with very short transmission distances. Recall that with a connector, a resonance can be set up with just a very small end-gap. Since intensity-modulated optical signals are not as coherent as RF signals, the relative impact is less, but it is still a concern.

One-way transmission devices called isolators are used at both RF and optical wavelengths to control reflections. Optical isolators are most commonly used to keep the output of SLM lasers stable. Reflected energy into

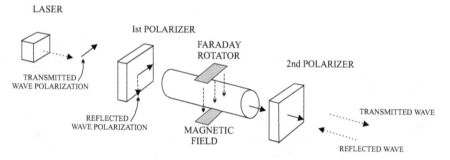

Figure 6-11 *SM isolator.*

a laser's cavity can disturb the output wavelength and add additional modes and noise.

An optical isolator is sketched in Figure 6-11. The parts are shown separately but when fabricated, the device is relatively small. In fact, SLM lasers can be purchased with the isolator in the packaged unit. An optical isolator relies on a magneto-optic effect first stated by Faraday in the 1800s: applying a steady magnetic field to certain substances will change their refractive indices. The three main parts of the optical isolator are the two polarizers and the (magneto-optic) Faraday rotator. A polarizer outputs the E-field component of an input wave so that it is oriented in the direction of the polarizer. An incoming wave that is circularly or randomly polarized will have only a portion of its energy passed. An incoming wave with linear polarization at 90° to the polarizer's alignment will be completely blocked. In Figure 6-11, a laser is shown as the source, with its output horizontally polarized. The radiation of semiconductor lasers is linearly polarized in a plane parallel to the plane of the active region. Since lasers have polarized outputs (Chapter 4), an isolator used in this application will cause very little insertion loss.

In the rotator, the input polarized field is rotated by an amount dependent on the rotator's internal substance, wavelength, interaction length, and field strength. Different substances can be used for the rotator medium. Quartz (crystallized S_iO_2) does not respond as well as zinc sulphide (Z_nS), for example. First, the input wave is polarized horizontally (Figure 6-11). The rotator is designed to add a 45° clockwise rotation at the design wavelength. The second polarizer is set 45° clockwise to the input wave, thus passing the output of the rotator. Any reflected wave will be polarized when it reaches the rotator. The rotator rotates the wave another 45° clockwise, putting its polarization at 90° to the first polarizer. This causes the reflected wave to be blocked.

6.8 Attenuators

Some optical detectors will be overloaded, perhaps even damaged, if the impinging signal is too strong. Optical attenuators are used to adjust signal levels at the input to an optical detector and also to control signals while making measurements. In testing a receiver, for example, an optical test signal level is lowered in known increments while the *signal-to-noise ratio* (S/N) or *bit error rate* (BER) is monitored. For APD receivers, there is an optimum input signal level that is first found through the use of a variable attenuator. Then, a fixed attenuator is placed before the detector to stabilize the receiver's performance. Like their RF counterparts, the connectors on an attenuator need to be compatible with the test equipment or device. Optical attenuators are available commercially for multi- and single-mode fibers with many different connector types. Attenuators are generally bi-directional.

Optical attenuators have internal components that provide the signal loss by either diverting or absorbing energy. Because losses can arise between two fibers in a number of ways, there are different ways of building an attenuator. A simple attenuator can be built by controlling the misalignment of two fibers. Another approach is to introduce an attenuating film between the two fibers. Yet another method introduces absorbing plates in a collimated beam. Grin-rod lenses are used on the input and output to collinate and focus the signal, respectively. A smaller version of this type of attenuator is obtained by using a mirror in place of the second lens.

If the internal alignments of the components causing the signal loss are fixed, the attenuator has a permanent loss value. The internal components or alignments can sometimes be made controllable, thus yielding a variable attenuator. For example, the lateral misalignment of two fibers can be varied, or an absorbing plate with varying width can be made to turn. Control can be achieved manually or remotely. Reflections are controlled in optical attenuators by angling the internal components. Calibration must be kept up-to-date. If attenuators are used in testing multimode systems, the fill condition of the input fiber must be understood by the user. It may be desirable to use a mode scrambler in the test setup. Recall that mode scramblers strip higher order modal energy and give a steady-state output.

6.9 Switches

A switch can be designed to be very fast or relatively slow depending on the circuit requirements. When a pulse is turned on/off, for example, the rise/fall times can be extremely fast. If a path for pulses or a signal is switched and held for an extended time, it is referred to as a *circuit switch*.

Current optical switches in use are circuit switches. At the present time, attempting to identify and re-rout individual bits of information at photonic wavelengths and Gigabit speeds is too demanding. Digital optical processing is still a hope for the future. Photonic packet switching, which will switch groups of optical pulses, is being heavily researched, however. The payoff in networking cost reduction and flexibility for even this first step toward true photonic switching is potentially large.

The optical switches used initially were electromechanically controlled. An example is sketched in Figure 6-12A. The signal from fiber 2 is switched to fiber 1 when the prism is repositioned. These were used for 1 × 2 protection switching, for example. Cascading these single-stage devices to obtain a larger I/O port count was an option but it led to degraded performance and higher cost.

An integrated-optics switch using electrical control is shown in Figure 6-12B. The basic technology used most often for these types of devices is Lithium Niobate with diffused titanium (Ti: Li Nb O_3). The titanium allows smaller waveguide sizes in the substrates, thus making single-mode devices possible. An applied electric field causes refractive index changes in each of the two waveguides, which then changes the input signal's phasing at the two output ports. An optical signal on fiber 1 can be made to appear on either fiber 3 or 4. These devices are very fast but also have significant insertion loss. Cascading them to make a larger switch matrix again leads to unacceptable signal degradations.

The availability of CWDM and DWDM and the need to differentiate between high-speed services at the optical level have brought about a major change in the application of optical fiber telecommunication systems. Wavelength multiplexing has greatly increased the number of optical channels on a single fiber. In addition, optical fiber components have been improved so all-optical networks (Chapter 8) can now be constructed. These two factors combined require that an optical, nonblocking, high-port count, low-cost "switch" matrix be available. The important element in these matrices is the switch, which is called an *optical crossconnect* (OXC). Most of the currently available optical crossconnects convert the optical signal to electri-

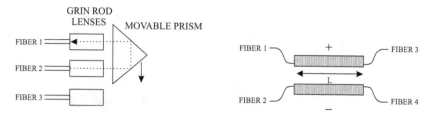

Figure 6-12 *Optical switches.*

cal, switch the electrical signal, and then reconvert to optical. This is called *OEO switching, (optical-electrical-optical)*. OEO switching is a costly approach, one reason being that the electrical switch matrix is dependent on the incoming data rate. A new technology, *MEMs* (*micro-electromechanical systems*), has emerged that will allow an all-optical, large port count, nonblocking crossconnect. This is called *OOO switching* (*optical-optical-optical*). The term *nonblocking* means that any input can be switched to any output, even though the fabric is being used by other signals. Further, the making of a new connection will not impact an existing connection.

Devices based on MEMs are very small because they are fabricated using existing semiconductor technology. An individual MEMs device is essentially a mechanical integrated circuit. One such device that can be used in optical crossconnects is a mirror. It can be gimbaled in multiple directions. The forces causing movement can be electrostatic, electromagnetic, or thermal. A matrix of these mirrors would make up the switch matrix. A nonblocking, 16×16, MEMs optical crossconnect switch is available commercially. The packaged crossconnect is about eight inches square and holds 256 individual MEMs devices. Available MEMs crossconnects are in two dimensions, but three-dimensional matrices are being researched.

6.10 Modulators

With high data rate modulation, the performance of injection laser diodes is adversely affected by the rapid changes in drive current brought on by the injected modulation. The index of refraction changes with the modulation. The carrier density then varies with time and a noise modulation called chirp becomes significant. Chirp broadens the optical carrier's spectrum just like shot noise does to an RF carrier. Since the effects of chromatic distortion are dependent on spectral width (Chapter 3), the end result is added signal degradation.

An important way of avoiding chirp is through the use of an external modulator placed between the laser and the fiber. The laser cavity is undisturbed, as long as reflections are low. A modulator can be constructed of III-V material and monolithically integrated with the laser. This approach is discussed more thoroughly in Chapter 8. Early modulators were made using integrated optical devices of titanium implanted into Lithium Niobate and were similar to the one pictured in Figure 6-12B. An electro-optic modulator will have only one input and one output port. The *Mach-Zender* modulator has two internal paths similar to the switch in Figure 6-12B. The phasing of the two internal paths is varied so that the output optical carrier is modulated. Phase modulation can also be achieved using these modulators.

7

Optical Fiber Systems

7.1   Introduction

Optical fibers provide the ability to use optics instead of metallic media to achieve numerous useful end results. In all cases the fiber requires an optical source and detector and cannot be used alone. As a result, optical fiber applications must always be considered in combination with their required E/O and O/E converters. This chapter looks at modern optical fiber telecommunications systems and the relationships between the source, fiber, and detector.

A general block diagram of a uni-directional optical fiber telecommunications system is shown in Figure 7-1. The optical source, LED or laser, is driven by the electrical baseband signal. Many different types of baseband signals are possible. The electrical section of the transmitter is designed specifically to accommodate the expected baseband signal and drive the chosen type of optical source. The electrical and optical sections of the transmitter are designed and built as one unit. There is not a one-size-fits-all transmitter available. Similarly, the receiver's electrical section is matched to the baseband signal and the chosen photodetector, PIN diode or APD. The transmitter and receiver are often combined in one unit, called a transceiver.

Terminal locations have a number of optical components. If only a simple link is used for the system, there might only be connectors, splices and an optical patch panel. Optical fibers now are used in complex optical networks that might require the addition of optical switches, multiplexers, demultiplexers, filters, waveguide converters, and routers at terminal locations. The

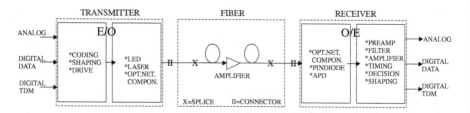

Figure 7-1 *Optical fiber system.*

fiber will be either multimode (MM) or single mode (SM), depending primarily on the required system length and bandwidth. Silica fibers are used almost exclusively for telecommunications applications. If the optical connection is long, optical amplifiers will be used at the transmitter, receiver, or at intermediate locations. Early systems used electrical regenerators that required O/E and E/O conversions. The availability of broadband optical amplifiers has lead to wavelength sharing of the fiber (WDM).

Figure 7-2 shows the receiver in more detail. Understanding the receiver and its effects on system noise are critical to understanding system performance. The choice of either a PIN diode or APD will determine the minimum received signal level and the type of noise that dominates in the receiver. The choice of electrical preamp is also an important consideration. The linear channel contains equalization, if needed, and amplification. Another important determination is the filtering of high-frequency noise. In a digital receiver, the filter also helps shape the incoming pulse for optimum detection. The digital electronics extract a timing signal and use it to strobe a decision on whether a pulse or no-pulse is present in a time slot. The detected pulses are reshaped to accommodate the electrical circuitry that follows in the signal path. In a regenerator this signal will drive the transmitting optical source.

Section 7.2 discusses the baseband signals carried on optical fibers and their performance requirements. Next, the possible signal degradations

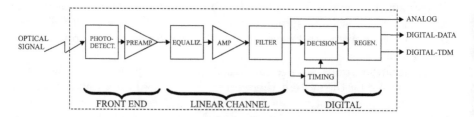

Figure 7-2 *Receiver block diagram.*

caused by the source and fiber are examined. Fiber nonlinearities and the transmission of solitons are also included. In Section 7.3, examples of two telecommunication systems are analyzed based on the performance requirements and noise/distortion contributors discussed in Section 7.2. These systems represent the extremes in possible applications: a very short, analog video system, and a long, high data rate digital system. Finally, field test equipment and measurement procedures are reviewed in Section 7.4.

7.2 System Signals, Noise, and Distortion

7.2.1 *Baseband Signals*
The two types of signals carried by optical fibers, analog and digital, have significantly different baseband (i.e., electrical) formats and spectra. The term *baseband* as used here refers to the signal prior to modulation on the optical source. In Figure 7-1, the baseband signal is input to the optical transmitter and output from the receiver. A baseband signal can be complete unto itself. It can also be obtained through combining or multiplexing other signals of lesser bandwidth/speed into a wider bandwidth signal. A video signal is complete but is still a combination of independent signals conveying the luminance, color, and sound information. An analog signal represents the originating electrical information in as exact a form as possible. Another way to think of an analog signal is that it is not a series of pulses. A digital signal is just that, on the other hand. It originates as either the output of a computing activity or as an encoded signal that began in analog format. Our voices begin in analog format, for example, but are transmitted in digital format. The longer distance portions of the national telecommunications network are almost 100 percent digital at this time.

The electrical format of these baseband signals is not of great interest for the discussion here. The spectral content, on the other hand, is important to optical transmission, since it dictates the required bandwidths and performance of components in the optical transmission path. Table 7-1 lists the most common baseband signals carried by optical systems, their bandwidths/speed, and some of their characteristics. The analog signals are described by their required bandwidths in MHz and the digital signals by their required bit rates in Mb/s.

Frequency division multiplexing (FDM) is achieved by placing separate channels adjacent to each other in the frequency domain. Each channel conveys independent information and uses some form of modulated carrier: vestigial sideband for TV, suppressed carrier AM for voice, FM for some multichannel CATV systems. Since optical fibers are being used, the baseband signal will have a wide spectral range. FDM for voice-band signals is

Table 7-1 *Optical systems baseband signals.*

Baseband Signal	Analog/ Digital (A,D)	Baseband Spectral Width (MHz or Mb/s)	Special Characteristics & Uses
NTSC Video	A	0.02-6	Security, local links
CATV	A	50-550	FDM-AM, >80 channels
CATV Multichannel FM	A	up to 300	FDM-FM, 30 MHz/channel
Antenna Remote	A	> 100	Radar, commercial AM/FM
Subcarrier Multiplexing (SCM)	A & D	up to 8	Microwave, FDM, Mixed services
High Def. TV	A	22	Entertainment, high resolution
Internet2	D	2,500	In development
Ethernet	D	1,000	10 Gb/s Ethernet to be adopted
SONET or SDH	D	up to 40,000	Time division multiplexed-TDM
Fiber Distributed Data Interface (FDDI)	D	100	200 km rings
Fibre Channel	D	1,060	Limited to 10 km

not listedbecause it is now obsolete. Voice-on-fiber is almost always transmitted using TDM, but improved Internet protocols will soon lead to expanded Voice-on-Internet usage.

Time division multiplexing (TDM) is achieved by first transforming independent signals into a digital format (A/D conversion), and then time interleaving their pulse streams into a faster pulse stream. Voice and data are combined to provide high data rate point-to-point connections. TDM was the first digital baseband signal used on optical fiber systems. Both North American and European TDM standards have been developed. In the 1980s, a great deal of effort was placed on developing international transmission standards for TDM. The *synchronous optical network* (SONET) standard was adopted for North America. SONET grew naturally from earlier digital networking TDM standards used for metallic/radio transmission. Outside the US, a different structure was adopted, the *synchronous digital hierarchy* (SDH). These recommendations, which evolved in the 1990s, have had a major influence on manufacturers and users of optical systems. Optical fibers have become so economically important that the communications industry has readily adopted these standards and recommendations.

The American SONET standards and the SDH recommendations have been brought into partial equality at high bit rates. The two standards allow optical interconnection beginning at the OC-3 (SONET) and STM-1 (SDH) rate of 155.52 Mb/s and extending in increments of 3 from the OC-3 level. The term OCs stand for *optical carrier* and STM stands for *synchronous transport module*. The highest level currently recommended is at 40 Gb/s (OC-768, STM-256). The OC-192, 10 Gb/s level is being installed extensively at this time. Manufacturers are beginning to make 40 Gb/s equipment available, either as single-rate transceivers or four coupled 10 Gb/s transceivers. As discussed later, the transmission problems become significantly more troublesome at very high bit rates. The OC-"N" and STM-"N" data signals are formatted using physical layer protocols intended for optical system transmission. The SONET standard distinguishes between an electrical drive signal (for example, STS-3) and the OC-3 optical signal. STS stands for *synchronous transport signal*. In these recommendations, fiber systems are categorized based on transmission distance (Table 7-2). The fibers currently recommended for these systems are:

- Graded index multimode in the 1300 nm window
- Standard single mode in the 1300 and 1550 nm windows
- Dispersion shifted single mode in the 1550 nm window

The SONET and SDH length descriptions are specific because the standards go beyond simply specifying data rates and give guidance on link design. The optical system descriptions provide objectives for fiber attenuation, optical source power output, and receiver sensitivity for systems up to 2.5 Gb/s. The maximum optical power per channel for WDM systems is also specified. Recommendations for 10 and 40 Gb/s systems are evolving. Receiver sensitivity is stated in the absence of external noise. The longer systems may use amplifiers and higher power lasers to exceed the listed distances. Manufacturers and service providers have developed the additional other terminology.

Local area networks (LAN) currently use primarily 100 Mb/s and Gigabit Ethernet. The 100 Mb/s signal can even be carried on good quality twisted pair for short distances. A *metropolitan area network* (MAN) is generally realized using SONET rings. Add/drops are currently achieved using electronics. All optical routing is anticipated as the next phase of improvement in the metropolitan core and access networks (Chapter 8). This increased capability is an important area of growth for optical equipment manufacturers and service providers. Future network growth will depend

Table 7-2 *Optical system description.*

Length or Coverage (km)	SONET Terminology	SDH Terminology	Other Terminology
< 2	Short-reach	Intra-office	Enterprise (Campus, Premise) Local Area Network-LAN
15	Intermediate-reach	Short-haul	Metropolitan Area Network-MAN (Metro-core and Metro-access)
80 @ 1550	Long-reach	Long-haul	Metropolitan Area Network-MAN
120 @ 1550		Very long-haul	Regional
160 @ 1550		Ultralong-haul	
1000			Long-haul
>1000			Ultralong-haul

largely on the growth of 10 Gb/s Ethernet and wavelength division multiplexing (WDM).

SONET based TDM transmission is not ideal for data traffic. In the 1980s, the need for a national and international network more compatible with data traffic emerged. More specifically, this issue of a telephony-based network not being satisfactory for data goes back to the anti-trust case against the integrated AT&T company in the late 1970s. The advent of packet switching for the Internet and widespread adoption of Ethernet answered some of the needs of data communicators. Currently, data traffic accounts for more than half of the total national traffic.

A protocol-based approach called *asynchronous transfer mode* (ATM) was developed in an attempt to meld data and telephony on the TDM network. ATM aggregates all end-user services into a single *user-to-network interface* (UNI) transmission signal for transmission at the SONET/SDH rates. The politically motivated need to keep local and inter-exchange business separate has led to ATM being used primarily at the local level. ATM is currently used extensively for *digital subscriber loop* (DSL) service. Data transmission systems such as *fiber distributed data interface* (FDDI), high-speed Ethernet, and fiber channel have been developed to answer some of the specific needs of data communication. The bandwidth capabilities of optical fiber and the bandwidth requirements of data signals have resulted in a natural partnership that is growing stronger and more important each year. The newest 10 Gb/s Ethernet standard is based on using only optical fibers. The choice of fiber depends on the distance and networking approach. Ethernet at 10 Gb/s will be compatible with SONET at the OC-192 level and will have a reach greater than 40 km on SM fiber instead of the current 5 km for 1 Gb/s Ethernet. This transport system alone is expected to have an important impact on the expansion of local and metropolitan area optical networks. Early installations might use four 2.5 Gb/s channels in a WDM configuration.

7.2.2 *Performance Objectives*

Each of these services places an end-user performance requirement on the optical transmission system. Many of these requirements are subjective; for example, most people have experienced noisy but still viewable video reception. If the noise is just thermal (white) noise, the tolerable S/N power ratio is different than if the noise is a single in-band RF interference. In addition, tolerance to impairments varies between individuals. Establishing end-user based requirements for each service and then using these to set optical receiver performance objectives would add an unnecessary level of complication to this discussion. Take for example a multi-channel CATV-FM system, where the signal energy in each channel is a fraction of the total average signal energy S. The noise per channel is a fraction of the total average noise (N). If the CATV signal were vestigial sideband AM instead of FM, a resulting S/N at the receiver's output would scale directly to each of the multiplexed channels. Since the per-channel signal is FM not AM, however, an additional analysis would be needed. The FM multi-channel format trades bandwidth for noise improvement. The frequency modulation increases the per-channel bandwidth requirement to 30 MHz from 6 MHz but decreases noise sensitivity by about 20 dB.

Because this type of complexity exists with many signals, objectives will be set for only the two basic categories of service, broadband analog and digital. For analog signals, a baseband S/N objective of 50 dB at the receiver output will be used. For digital signals, a baseband bit error rate (BER) objective of 10^{-11} will be used, which is the SONET/STM objective for line rates above 1 Gb/s.

A BER describes the quality of a digital signal by stating an estimate of the average probability of an error in a stream of bits. A BER of 10^{-6}, for example, means that one bit in a million is likely to be in error. A digitized voice conversation will be bothered very little by a BER of 10^{-6}. Data traffic requires BERs in the region of 10^{-10}. Extra bits are often added to the digital pulse stream to help detect and correct errors at the receiver. This technique is called *forward error correction* (FEC) and is used extensively on submarine fiber optic systems.

Meeting these objectives requires that a number of noise contributors be considered during the design process. They must also be measured and kept under control during service. The transmitter-receiver combination will have *intersymbol interference* (ISI) and a non-zero BER. It is caused by the receivers shot noise, dark current noise, thermal noise, and filter shape. The S/N performance of the receiver was calculated based on these noises. To these four must be added the possible effects of other noise sources and distortions: dispersion, reflections, laser intensity noise, timing jitter, crosstalk, mode partitioning, modal noise, polarization mode dispersion, chirp, nonlinear distortions, and an accommodation for aging. In both the analog and digital case, these additional impairments require that the system be designed for a higher signal level (S), than that found to just meet the transceiver requirements. A power margin, discussed later, is designed into the system to achieve this objective.

Since optical fibers almost always use digital modulation, a more detailed analysis of the BER is needed. Figure 7-3 plots representative BER curves as a function of the receiver's average S/N. Curve A gives the result when only white noise is present within the signal band. This is referred to as the *error-function* (erf) curve. The assumptions that yield this curve are: ones and zeros equally probable, noise is equal for ones and zeros, no light when a zero is present, decision threshold at half the peak power of a one,

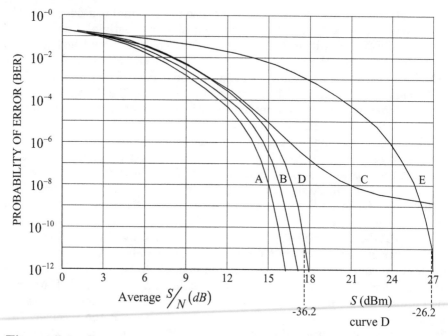

Figure 7-3 *Bit error rate curves.*

no intersymbol interference. Note that if the S/N becomes low enough, many bits of information are lost. When the S/N is very high, there are essentially zero errors. A BER vs. S/N curve usually has a very steep slope for high S/N. A small increase in N or decrease in S brings about a significant deterioration in the BER. If this were a plot of digital radio performance, a carrier/noise (C/N) ratio would be used. Conventional radio modulation places most of the energy in the carrier and very little in the sidebands. Even with spread spectrum techniques, the carrier energy is still the signal energy.

Curve B shows the effects of transceiver degradations. The back-to-back transceiver performance can be derived early in the system design or during installation. Note that at S/N = 17 dB, if the signal level were increased by 1 dB, a 10^{-11} BER would again be achieved. This is called a 1 dB power penalty. SONET/STM recommendations allow a 1 dB power penalty for the transceiver alone. Curve C illustrates the effects of some of the line and system degradations, which are discussed more thoroughly in the following sections. The tail on curve C would be caused by a degradation that was independent of signal power but becomes more controlling relative to other degradations as S increases. An example of this would be mode partition noise caused by the laser source.

Curves A–C are general, not based on specific system parameters. Curve D corresponds to the ISI penalty of 2 dB accepted for the model system's transceiver. The eye closure is 37%, and is caused by the responses of the laser, fiber, receiver together. The receiver filter is assumed optimized to give a raised cosine response for the channel in the time domain.

Curve E gives an estimate of the BER for the system modeled in Section 7.3.1. When the signal at the detector has dropped to its lowest acceptable level only receiver and optical amplifier noises (ASE) are applicable. Based on the receiver noise alone, the sensitivity is −36.2 dBm. The ASE noise alone causes a 10 dB power penalty. Adding the ASE noise means that the receiver input cannot be below −26.2 dBm if the 10^{-11} BER objective is to be met. The amplifiers provide sufficient power margin to overcome the ASE noise. Additional noise and distortion sources will take away still more of that margin.

7.2.3 Digital Line Coding

Digital signals are coded before transmission. The baseband signal might already have protocol information in it. The main reason for using line codes is to minimize errors upon reception and decoding. The coding matches the digital signal to the specific needs and characteristics of the transmission system in an attempt to optimize quality. With a microwave radio system, for example, 16-phase shift keyed (16 PSK) modulation might be used so that stringent FCC bandwidth usage requirements can be met. The T-1 paired cable system uses a bipolar, or alternate mark inversion line format to minimize low-frequency energy in the signal spectrum and maximize timing energy. The choice of line code determines a number of system characteristics: minimum bandwidth required for transmission, ease of clock extraction, ease of decoding, low frequency energy content in the baseband spectrum, ease of detecting errors. Because optical fibers have almost unlimited bandwidths, the choice of line codes is less restrictive but still an important design parameter. Analog systems use pre-emphasis/de-emphasis and equalization filtering to help match the signal to the system.

Most digital optical systems use either of the two line codes illustrated in Figure 7-4: nonreturn-to-zero (NRZ) and return-to-zero (RZ). The NRZ format is preferred at high bit rates because it requires a smaller baseband bandwidth.These codes are in binary (two-level) transmission format. There is little need to save bandwidth with optical fiber systems by using multi-level codes. Multilevel codes make the baseband electronics more complicated and costly. With binary coding, a bit of information is sent in one time slot, T. The bit rate is the information transfer rate for a digital system. A zero, or lack of a 1, is also useful information when decoded so the

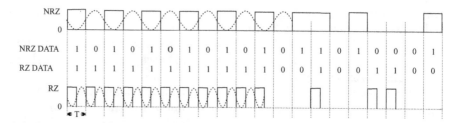

Figure 7-4 *Optical binary line codes.*

time slot rate is the bit rate. Another data term often seen is the *baud rate*, which by exact definition is the number of signal transitions per second. With binary transmission the rate of bit transmission is the baud rate. The term *symbol rate* describes how fast symbols are sent. Optical systems use binary transmission almost exclusively so all these rates are equal; thus, the distinction between bits and bauds becomes irrelevant.

The NRZ code in Figure 7-4 requires the smaller baseband spectral width of the two binary line codes. One disadvantage with NRZ is that long strings of ones can generate significant low frequency energy, leading to a degradation called *baseline wander*. Another disadvantage of NRZ is the lack of timing information for the receiver's decision circuits. There are a limited number of transitions at the line rate (T) in the pulse stream. With NRZ, timing information is sometimes added to the pulse stream or sent separately. Also, scrambling is sometimes used to change an all "1s" sequence into a random sequence of "1s" and "0s"

The RZ code in Figure 7-4 includes significant timing information within the information stream. There are many transitions at the bit rate. Given the same bit rate, 1/T, the RZ code makes 1-0 and 0-1 transitions twice as fast. Indeed, one variation of RZ coding, called Manchester coding, uses a transition every time slot. A negative-going transition can indicate a "1" and a positive-going transition a "0." When ones or zeros occur in groups the coder resets at the end of a bit interval. Figure 7-4 shows RZ using a 50 percent duty cycle pulse. This is the pulse width normally used for lower data rate systems. Soliton pulses, discussed later, use RZ pulses with a very small duty cycle. Soliton transmission requires very short, high-intensity pulse shapes to take advantage of the fiber nonlinearities.

With a well-designed receiver, RZ will have a slight advantage over NRZ. The receiver has to have wide enough bandwidth to accommodate the RZ code without excess distortion. This result assumes average power, not peak power in calculating S/N. Using peak power instead of average power to obtain the S/N reverses this advantage. Since lasers actually operate somewhere between peak and average power on an instantaneous basis, neither RZ nor NRZ has a distinct noise advantage at the receiver.

The system bandwidth needed to transmit pulses satisfactorily is directly related to the pulse repetition rate and the rate of pulse transitions. As with all transmission systems, a compromise has to be reached between the pulse's bandwidth requirements and the system's available bandwidth.

A long series of alternating ones and zeros in the NRZ format is shown in Figure 7-4. If a Fourier analysis were applied to this square wave, it would show spectral components at the frequency 1/(2T) and all odd harmonics. A sine wave of frequency 1/(2T) is overlaid on the alternating sequence. Since all systems, even optical systems, are bandwidth limited, the higher harmonics will be greatly attenuated. The pulse out of the receiver's linear channel will have rounded edges and look very much like the sine wave at frequency 1/(2T). There would also be a DC component that would eventually be blocked by AC coupling in the receiver electronics. The DC has to be restored in the decision and reshaping/regeneration circuitry that follows the linear channel.

A digital signal is a random string of pulses and seldom a long string of ones and zeros. Mathematical analysis shows the spectrum of random pulses to have significant content at low frequencies. This is logical since, with random pulses, there are many zeros possible between the ones. A baseband spectral analysis for this random signal would show an envelope, tapering to 3 dB down at 1/(2T) and decreasing gradually to zero at very high frequencies. As a first approximation, the bandwidth of the NRZ coded system has to pass frequencies up to at least this half bit rate. This frequency [1/(2T)], will be used to establish the required receiver bandwidth for NRZ coded signals. Filters are used in the receiver to limit out-of-band noise and shape the pulse for optimum detection. The filter bandwidths assumed are equal to the bit rate. This is wide enough to avoid introducing excess ISI yet narrow enough to limit noise for both NRZ and RZ pulses.

From Chapter 3 the system's 3 dB frequencies and the resulting pulse spread are related:

$$\Delta\tau = \frac{1}{2\sqrt{2}f_{3dBelectrical}} \qquad 7.1$$

$$f_{3dB\text{-}optical} = \sqrt{2}f_{3dB\text{-}electrical} = \frac{1}{2\Delta\tau} = \frac{0.5}{\Delta\tau}$$

Assuming an NRZ signal with pulse period T and a 3 dB electrical bandwidth of 1/(2T), the resulting time spread at the full-width-half-max (FWHM) points becomes:

$$\Delta\tau_{NRZ} = \frac{T}{\sqrt{2}} = 0.707\ T \qquad 7.2$$

Equation 7.2 basically translates a time delay imposed on the source wavelengths by the fiber into a bandwidth limitation that determines what bit rate can be transmitted. Applying equation 7.2 allows a derivation of a possible bit rate objective (1/T) if the fiber delay ($\Delta\tau$) is given, or a derivation of a delay objective if the bit rate is given. This approach is valid only for the fiber. All components, even optical fibers, have bandwidth limitations that can be a factor if a signal contains high frequencies in its spectrum. These bandwidth restrictions result in a rounding of the edges of a pulse. One of the basic assumptions behind the derivation of equation 7.2 was that an electrical bandwidth of B/2 (B=NRZ bit rate) is sufficient to pass the digital pulse stream. The pulse edges will be rounded but, when many pulses are viewed together on a scope, there will be a clear center section or *eye*. A decision made at the time of the maximum eye opening will result in almost error-free transmission.

Delays are not computed for the transmitter and receiver, E/O and O/E devices. Instead, their 10–90% rise time responses to a step input are used to describe their time (bandwidth) performance. The rise times of the three main system components are combined to obtain a system bandwidth performance estimate. The fiber rise time is obtained from equation 3.4 (fiber rise time = $\Delta\tau$). This approach to bandwidth analysis basically assumes the time domain responses of each of the components is exponential.

The results in 7.1 and 7.2 are based on an intuitive approach and are not mathematically rigorous. Another approach to the translation of dispersion-caused spread to allowable bit rate assumes Gaussian input and output pulse shapes in the time domain. The output pulse shape will have a spread caused by the three dispersion effects: material, waveguide and modal. The analysis using this approach is too detailed for our purposes, but the result is important because it shows a similar relationship to that in Equation 7.1. With Gaussian pulse shapes, the 3 dB optical frequency is:

$$f_{3dB\text{-}optical} = \frac{0.44}{\Delta\tau} \qquad 7.3$$

The numerator of 0.44 compares closely to the 0.5 numerator obtained using the intuitive method above. An advantage to using the Gaussian analysis is that ISI and the resulting power penalty can be predicted with more mathematical rigor. If the fiber has a 3 dB bandwidth greater than or equal to $0.44/\Delta\tau$, a transmitted bit rate equal to this frequency will have <1 dB ISI power penalty. This is because 95 percent of the pulse's energy lies within the time span of T. Using the 0.5 factor results in a 2 dB power penalty, again assuming Gaussian pulse shapes. In the equation 7.2 relationship, the 0.707 multiplier becomes 0.6, thus decreasing the allowable bit rate for a given value of fiber dispersion.

As might be expected, a similar analysis of the RZ format using the long string of alternating ones and zeros results in a wider electrical bandwidth requirement, 1/T. This is logical since the transitions are occurring at twice the rate of NRZ transitions. The allowable RZ pulse spread relationship to the system bandwidth and bit rate becomes:

$$\Delta\tau_{RZ} = \frac{1}{2\sqrt{2}f_{3dB\ electrical}} = \frac{T}{2\sqrt{2}} = 0.35\ T \qquad\qquad 7.4$$

This is half of the allowable total spread of NRZ coding given the same bit timing T. The RZ pulse will spread the same relative amount as the NRZ pulse, but the pulses will be occurring at twice the rate. The wider bandwidth requirement for RZ causes a moderately increased bandwidth requirement for the receiver. Receiver design is helped because of the RZ timing advantages.

The NRZ or RZ code is chosen to optimize transmission given the optical system characteristics and service requirements. Other forms of line codes are used in addition to NRZ or RZ to give added protection from errors. These basically increase the redundancy within the data stream. Generally, this is accomplished by adding bits to the original signal. This coding is placed in the data stream before it is input to the optical source. After demodulation and processing, the added redundancy is used to check whether the received data is in error and then to correct it if necessary. Examples of this coding are found in the 4B5B block code used with FDDI and the 8B10B coding of fiber channel. An mBnB block code translates m information bits into n binary bits for transmission, thus adding the required redundancy. The required system bandwidth is increased by n/m, but timing, in-service monitoring, and baseline wander can be improved. Another type of baseband coding, *forward error correction* (FEC), uses cyclic codes with added redundancy. A 216 bit to 224 bit coding has been shown to improve the BER by a factor of about 10^3. An older technique called *automatic repeat request* (ARQ) is also used occasionally, but causes delay problems for voice and video signals because real time transmission is required in two directions.

7.2.4 Noise, Distortion, and Nonlinearities

The initial analysis of receiver performance in Chapter 5 discussed dark current, shot, and thermal noise. Preamplifier noise was approximated using an added noise figure on the load resistor's thermal noise. Surface current noise in the diode was not considered since it makes a very small contribution in high quality photodiodes. Both PIN diode and APD receivers were analyzed. This section discusses the additional noise and distortion problems that can degrade optical fiber system performance. These degradations are combined with the receiver noise in the system analysis of section 7-3. Finally, a discussion of both the problems and advantages posed by fiber nonlinearity is provided.

In an optical analog receiver, a signal meets transmission objectives by having an adequate S/N. This is achieved through having a high input signal, low noise, or a combination of the two. If the bandwidth is severely limited, there will be an added problem of distortion. Harmonic generation caused by nonlinearity in the transmission path can cause intermodulation interference. These add to the system shot and thermal noise. Also, crosstalk from other signals can add to the noise if it is not adequately filtered.

In an optical digital receiver, a signal meets objectives if the ones or zeros are accurately detected. Figure 7-2 shows in block diagram form the receiver's electrical section, which includes the linear channel amplifier/filter, timing clock extractor, decision circuit, and pulse regenerator. The linear channel amplifies the signal, filters out higher frequency noise, and shapes the pulse to optimize decision efficiency. Generally, *automatic gain control* (AGC) is used to avoid receiver saturation. If the filter bandwidth is made large to limit additional intersymbol interference (ISI) and widen the eye, the receiver noise will be high. If the filter bandwidth is made narrow to lower the amount of receiver noise, the eye is squeezed by ISI. The best system pulse output, with filter, has a raised cosine impulse response. If only receiver noise is present, the raised cosine filter provides a good compromise and is generally adopted by manufacturers.

Figure 7-5 illustrates what appears on an oscilloscope when many received NRZ pulses are overlaid on an oscilloscope screen. This is called an *eye diagram,* and is very useful in quickly judging the quality of pulse transmission (Section 7.4). An RZ eye diagram would not have lines across the top 1 level. The sweep is synchronized to the bit rate. The receiver's decision circuitry is designed to make a 1/0 determination in the middle of the eye. The more open the eye, the less likely a mistake will be made in

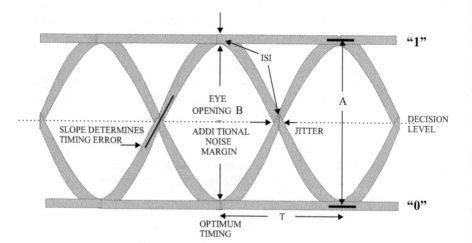

Figure 7-5 *NRZ eye diagram.*

this determination and the better the BER. The shaded area is intended to show the smear caused by many pulses overlaid on one another. The eye diagram smear shows that the pulses are spread and slightly offset relative to each other in amplitude and timing. Templates are commonly used in the field to help measure whether an eye diagram shows satisfactory system performance. The performance is unsatisfactory if pulse traces cross inside the template's eye. If a BER is very low, the eye diagram will not show much degradation and a BER measurement might be needed (Section 7.4).

The lateral offset is the *timing jitter*. The timing clock or strobe is usually extracted from the incoming signal and determines the time when a decision should be made on whether a 1 or 0 is present. Additional energy at the timing frequency may be added at the transmitter to ensure a reliable timing signal. There will be some noise on the top and bottom, 1 and 0 levels. This noise, which may be asymmetrical, comes from the receiver and some of the sources discussed below. In an APD, shot noise will be more significant than thermal noise, which will make the top 1 noisier. Also, lasers are normally operated with their bias current just above the threshold current to avoid excessive mode-partitioning. As a result, light will be received during a 0 because the optical source is not completely turned off and the eye will again have asymmetrical noise. The ratio of the optical energy emitted with a 1 to that emitted with a 0 is called the *extinction ratio*. Extinction ratios are generally 10 or more. Occasionally, the extinction ratio is stated in the reverse, which means it would fall in the range 0 to 1. The decision circuit has a threshold level set approximately in the middle of the pulse/no-pulse, 1/0, levels. Sometimes the decision level requires adjustment because of the presence of asymmetrical noise.

Finally, the pulses will be spread because of dispersion and component bandwidth limitations. The spreading causes ISI, which acts to narrow the eye both horizontally and vertically. All of the degradations tend to close the eye down, which then causes bit errors. The eye does not have to be completely closed to have a bad BER. The noise riding on the pulses is statistical in nature. A noise spike occurring every million bits is difficult to see on an eye diagram but is sufficient to make an intolerably high BER for data signals.

When BER curves are used to assess digital system performance, an offset in the curves occurs when significant noise producers are added to the normal receiver noise (Figure 7-3). The power margin makes it possible to operate the system at the desired BER in spite of these added noises. If a BER measurement shows a 10 dB offset, for example, the system needs at least that much extra margin to keep the received signal sufficiently high and the BER low. The 10 dB offset is a power penalty.

If an eye is very open, it can be assumed that the system is operating at a high S, has few distortions and adequate margin. It is helpful to be able to relate a power margin required on a BER curve to the eye diagram, since the eye diagram is much easier to obtain in the field. The additional power margin required to maintain a desired eye opening can be approximated using the following equation:

$$E_{+MAR.} = 10 \log_{10} \frac{A}{B} = 10 \log_{10} \left(\frac{1 - D_i}{1 - D_i - D_S} \right) \text{dB} \qquad 7.5$$

where D_i = initial eye degradation above sensitivity limit, D_s = subsequent degradation. The power relationship at the optical level between pulses translates directly to the eye diagram because the optical power generates a current which translates to a voltage.

The 2.5 Gb/s system designed in section 7.3.2 is dispersion limited. Using equation 7.2, a 2 dB power penalty results from the amount of spreading allowed in the initial system design objective (rise time objective = $\Delta\tau$ = 280 ps). This includes the transceiver response times. Starting with an initial eye degradation of zero, Equation 7.5 estimates that the eye will be 63 percent open if only the 2 dB penalty is present. This seems to be a severe closure except for the fact that the major contributors to eye degradation are included in the 37 percent closure. Further, the remaining contributors are largely under the control of the designer. If the final eye opening is to be held to 60 percent, for example, Equation 7.5 estimates a power margin of 13.2 dB is required for any additional noise/distortion. The model system design derived in Section 7.3.1 has a power margin of 20.4 dB available even after amplifier noise is taken into account. This is more than sufficient to meet objectives.

7.2.4.1 *Intersymbol Interference (ISI) and Dispersion*

The total time spread experienced by the signal is caused by the finite speed of response of the transmitter and receiver, as well as the dispersion in the fiber. The fiber can have modal, chromatic or polarization-mode dispersion. The effects of dispersion on the signals are calculated using pulse rise times (Chapter 3). The method applies to both analog and digital transmission. If the total system spread is too high, the pulse will distort significantly. With analog transmission, too much pulse spreading means the system bandwidth is severely limited. An analog signal will experience harmonic distortion and intermodulation when the signal bandwidth is restricted. A digital signal will experience intersymbol interference (ISI) and a resulting high BER. ISI is caused by:

- inadequate transmitter, detector or pre-amplifier bandwidth
- improper receiver filtering and equalization
- inadequate low frequency response which leads to baseline wander
- various distortions caused by fiber dispersion

The eye diagram in Figure 7-5 illustrates ISI. Either rise time or bandwidth specifications can be used to calculate the effects of pulse spreading. Component and device data sheets will give one or the other. The 35 percent RZ and 70 percent NRZ pulse spread objectives at the FWHM pulse times were derived in Section 7.2.3. They result in a small power penalty but an apparently significant eye closure. The ISI resulting from just meeting the 35/70 percent RZ/NRZ pulse spread objectives results in a 2 dB power penalty and a 37 percent eye closure. A mathematical analysis supporting this translation would be too detailed for this treatment. Other possible contributors to ISI are echoes, mode partitioning, chirp, and polarization mode dispersion. These are guarded against, respectively, by using very good connectors/splices, higher laser bias, external modulation, and high quality fibers.

7.2.4.2 *Optical Source Induced Noise*

The optical source can be a direct or indirect contributor to system noise and distortion. Mode partitioning noise would be an example of a direct contribution and time spread caused by chromatic dispersion an indirect example. The optical source, LED or laser, can contribute harmonic and intermodulation distortion to an analog signal. In addition, an LED will be bandwidth limited because of its relatively slow internal response. With a digital signal, the LED's slow response will contribute to ISI. Unlike a laser, the LED will not contribute internally generated noise, however. The laser has a number of potential noise sources, whose importance in a system design depends primarily on how the laser is biased and how the light output is modulated. An internally modulated laser will contribute more noise/distortion than an externally modulated laser because its index of refraction is changed by the modulating energy. A laser diode source provides higher power and faster response (higher bandwidth) but, in return, requires attention to additional noise sources.

Recall that light emitting diodes produce a broad spectrum, noncoherent output and have no threshold. An ILD has a narrow spectrum, a relatively coherent output, and a threshold drive current that must be met or exceeded to sustain the oscillation. An LED can be turned on from zero drive current, but the ILD must have a bias. If a drive current is below the threshold current, the light is spontaneous. If above the threshold, the laser is emitting stimulated light. The modulation drive currents add to the bias current, but the total drive current is always kept above the threshold current. As a result, a small amount of light will be generated in the 0 state. In Figure 7-5, the 0 level will have added noise, which causes an extinction ratio noise penalty. If the extinction ratio is greater than 10 dB, the added system penalty will be less than a dB. There are a number of advantages to accepting this small penalty. A bias above threshold gives faster on/off times. It also reduces laser-induced relative intensity noise, reflection noise, mode partitioning, chirp, and modal noise.

Relative intensity noise (RIN) is caused by random fluctuations in a semiconductor laser's output intensity. RIN is not present in LEDs and is usually only a problem with analog transmission. The noise occurs because the light is being quantized into photons with some degree of randomness. The noise is present even under the CW operation used with an external modulator. The lower the bias current relative to the threshold current, the higher the RIN. Biasing too close to threshold should be avoided if RIN is a potential problem. In addition, when the driving signal's bandwidth approaches the laser's resonance frequency (Chapter 5), the RIN noise is increased. The 1550 nm InGaAs-DFB laser used in the Section 7.3.2 model system analysis is assumed to have RIN of approximately –150 dB/Hz. The RIN contribution to the receiver's total noise current is calculated as follows:

$$N_{RIN} = RIN \, (RP_{in})^2 \cdot B \cdot R_T = 1.5(10)^{-13} \, A \qquad 7.6$$

where RIN = –150 dB/Hz, R = responsivity = 1, B = $2.5(10)^{-9}$ = bandwidth, P_{in} = receiver sensitivity (–35.4 dBm), B = bandwidth, R_T = detector load resistor = $(10)^5$ ohms. This noise level is not significant relative to the shot noise in the APD receiver used in the digital model system. Because RIN varies with the square of the input power, the RIN contribution increases with higher receiver input power. An analog system requires higher power levels to achieve a satisfactory S/N. If an analog system uses a laser source, RIN has to be considered as a major noise contributor.

Reflection noise is induced in a laser when energy reflected back into the oscillating cavity affects the laser's performance. The reflections increase the RIN. A feedback ratio of less than 50 dB can increase RIN by 10 to 20 dB. A reflection of –10 dB can cause a power penalty greater than 4 dB for a high data rate digital system. This is the level of reflection an air-gap connector would have, for example (Chapter 6). Reflection noise will be relatively more significant in single-mode fiber systems using SLM lasers to reduce dispersion effects. Increasing the signal level does not gain protection from reflection noise. The importance of reflection noise relative to the other noise sources increases as the signal increases. Curve C in Figure 7-2 shows a tail at high S/N. Reflections are one of a number of ways this tail can be generated. Reflection levels can be decreased at connectors in a several ways: index matching fluid, angling the fiber surfaces, and using physical contact connectors. Optical isolators also are used to reduce reflections. These provide >25 dB return loss.

In addition to reflection noise, two reflections can generate an interfering echo at the receiver that causes ISI. The small distance between connector faces can set up a resonant cavity if the optical source is sufficiently coherent. A DFB laser with very small linewidth (for example, 0.0004 nm or 50 MHz at 1550 nm) will have a coherence length of about 2 meters (Chapter 4). A jumper of that length with poor return loss at the two connectors can generate a significant amount of ISI.

Mode partition noise (MPN) occurs when multiple longitudinal modes in a laser's output have amplitude fluctuations. These amplitude changes occur primarily when pulses are turned on and off. The total output power in the pulse grows smoothly in accordance with the power output-current input characteristic (Figure 4-9) but the distribution between modes changes. The fiber's chromatic dispersion will transform these amplitude variations into MPN. The ratio of the main mode's average energy to the strongest side mode's average energy is called the *side-mode suppression ratio*. A side-mode suppression ration > 50 dB is generally required to keep MPN acceptably low. Similar to reflection noise, the relative contribution of MPN increases with the signal level and will show up as a tail on the BER curve (Figure 7-3). An SLM laser or external modulator can be used to reduce MPN. Also, if the laser bias level relative to the threshold is high, the MPN is lessened because the laser's spectrum reaches a stable condition faster. Of course, a higher bias will result in an increase in extinction ratio noise.

Chirp is the dynamic spectral broadening of an injection modulated SLM laser's output when it is turned on. The broadening of the laser linewidth increases pulse spreading and, ultimately, ISI. The modulation current induces changes in the carrier density within the semiconductor. The chirp can be in the range of several tenths of a nanometer. Chirp is a serious noise source for high data rate systems operating at 1550 nm on conventional SM fiber. If the extinction ratio is high, i.e., the bias is close to the threshold, the chirp penalty could be a few dB. Raising the bias reduces the chirp but increases the extinction ratio penalty. The potential effects of chirp can be reduced by using dispersion shifted fiber or fiber dispersion compensation (Chapter 8).

Modal noise occurs when the modes in a multimode fiber experience differential phase and attenuation during transmission. Light from a coherent source (laser) launched into a multimode fiber will generate a speckle pattern. The speckle pattern results from interference and reinforcement between propagating modes. Differential changes in phase and attenuation between the modes, at lengths shorter than the source coherence length, leads to speckle pattern changes. Upon detection these changes, which can occur at connectors, splices, and microbends, can become modal noise. Use of a low coherence, multimode laser minimizes modal noise by decreasing coherence length and reducing speckle pattern contrast. Single-mode fibers might have a modal noise problem if the operating wavelength is close to the fiber's cutoff wavelength (Chapter 3). A short jumper made of the same fiber might sustain the next higher order mode and thereby cause modal noise. Using jumpers with cutoff wavelengths higher than the fiber, cutoff wavelength will reduce this possibility. Also, longer jumpers can give reduced speckle pattern contrast and less modal noise.

7.2.4.3 *Optical Amplifier Noise*

Optical amplifiers generate *amplified spontaneous emission* (ASE) noise. The noise arises from the spontaneous recombination of electrons and holes in the amplifier. A noise figure per amplifier will be in the range of 4 to 6 dB. Recall that the noise figure is equal to the ratio of the S/N before amplification to the S/N after amplification. The ASE noise builds up with each successive amplified fiber section. It is slightly higher in EDFAs that are counter-pumped as opposed to those forward-pumped. Also, the noise build-up is faster in cascaded amplifiers that have higher gain. A power penalty due to this S/N deterioration must be accounted for in the power budget calculations. For cascaded amplifiers with equal gain sections, the S/N can be estimated from

$$\left(\frac{S}{N}\right)_{ASE} = \frac{P_T \cdot 10^{-(AL/10)}}{N \cdot K \cdot hf \cdot B} \qquad 7.7$$

where P_T = transmitted power, A = attenuation in dB/km, L = fiber length per amplifier section, N = number of equal length amplifier sections, K = amplifier noise figure, hf = energy per photon, B = bandwidth.

Using the model system developed in Section 7.3.2 (Amp.NF = 6 dB = 4, 10 sections at 100km, 0.25 dB/km attenuation, B = 2.5 GHz):

$$\left(\frac{S}{N}\right)_{ASE\text{-}MODEL} = \frac{10^{-3}10^{-(0.25\cdot100/10)}}{10 \cdot 4 \cdot 1.3(10^{-19}) \cdot 2.5(10^9)} = 243 = 23.8 \text{ DB} \qquad 7.8$$

Without the added ASE noise, the S/N (based on average powers) required to achieve a system objective of 10^{-11} BER is 17.4 dB (Figure 7-3). The digital system model uses an APD-TIA receiver, and the 17.4 dB S/N objective is met at a receiver input power of –36.2 dBm (section 7.3.2). The ASE noise is 6.4 dB below the receiver noise for this example (23.8–17.4). The receiver sensitivity level of –36.2 dBm is never reached in an amplified system because the received powers are always much higher. As a result, the amplified system can offer a very good power margin based only on receiver noise: 30.4 dB for the amplified system in Section 7.3.2.1. This very large degree of protection is needed because of the ASE noise. If all amplified sections are equal, the system penalty is given by:

$$F_{ASE \text{ penalty}}(G) = \frac{1}{G}\left(\frac{G-1}{\ln G}\right)^2 \qquad 7.9$$

Where G = total section gain = 25 dB = 316. The penalty for the system model is calculated as 10 dB, which will use a significant portion of the margin.

If a longer system is desired, the number of amplifier sections can be increased. If all the system noise were allocated to ASE noise, the ASE S/N would be 16.4 dB, instead of the 23.8 dB result in Equation 7.8. Using the parameters of the model system, this would yield a system length of about 5700 km. This approach is plotted in Figure 7-8 with bit rate as an independent variable. This design would not be useful, or even possible. There are other system distortions and noises that might degrade performance and use up an available power margin.

7.2.4.4 *Polarization Mode Dispersion (PMD)*

Polarization mode dispersion is a serious concern in the design and operation of 10 and 40 Gb/s, SM systems. Lower speed and analog systems are generally not bothered by PMD. For example, the 2.5 Gb/s system analyzed in Section 7.3.2 has PMD that is not significant relative to other dispersion effects. In a single-mode fiber the electric field has two modes, orthogonally polarized to each other (Chapter 2). In ideal fibers, with perfect circular and rotational symmetry, the two modes will propagate unchanged. Actual fibers have imperfections that result in the two modes having different phase velocities and different effective refractive indices. The difference in the effective refractive indices results in a difference in propagation time and, as a result, pulse spreading. This is called fiber birefringence. PMD is not a concern in multimode fibers.

The effect of fiber PMD on system performance is estimated by

$$(\Delta \tau_{PDM}) \text{ expected value of spread} = D_{PDM} \sqrt{L} \qquad 7.10$$

where D_{PDM} = *average PMD parameter* = $(0.1 \rightarrow 1)$/ps/$\sqrt{km}$, L = total fiber length. Fiber quality and cable environment determine the PMD parameter value. The instantaneous value of PMD in an operating system will vary, but the long-term average used in Equation 7.10 is fairly constant. For the 1550 nm digital system model in Section 7.3.2 (DFB laser, $\Delta \lambda = 0.1$ nm, 1000 km, dispersion of 2 ps/nm-km), assuming a PMD value of 0.5 ps/$\sqrt{km}$), the PMD spread is 15.7 ps. This amount of additional spread is not significant for the 2.5 Gb/s system analyzed in section 7.3.2. The effect would be much greater for a 10 Gb/s system, however, because the tolerance to dispersion decreases with increasing bit rate. A 14 ps spread at 10 Gb/s would result in a power penalty of about 3 dB. The next generation, 40 Gb/s systems will probably need total PMD to be < 3 ps. One survey estimated that 20 percent of currently installed SM links >300 km are unsuitable for 10 Gb/s transmission because of PMD. That survey also found 75 percent of the links inadequate for 40 Gb/s.

PMD is unavoidable in fibers but can be kept low (<0.2 ps/$\sqrt{\text{km}}$) by controlling fiber concentricity and rotational uniformity. This is being achieved in current manufactured fibers. Special polarization maintaining fibers can also be used (Chapter 3), but they are expensive and require optical components capable of maintaining polarization. Lumped PMD is also found in the passive devices used in many SM optical networks. Fixed compensation is available to correct this type of PMD. Dynamic equalizers will be required to compensate for the random PMD caused by a fiber in an operating environment.

Another problem associated with PMD is *polarization dependent loss* (PDL). It is defined as the variation in power at the output of a device as the input signal's state of polarization changes. A linear polarizer is designed to stop all but the selected *state of polarization* (SOP). Its PDL will be > 30 dB. All other optical components have small, but sometimes significant PDL. An isolator has a PDL of < 0.3 dB. An angled PC connector will be < 0.1 dB. The state of polarization (SOP) in SM fiber systems is not controlled. Since intensity modulated systems do not require a stable SOP, PDL will cause a small amount of added, unpredictable loss. As a result, a loss allocation should be considered during the design phase, and PDL assumed to be present during measurement and turn-up.

7.2.4.5 Nonlinear Fiber Performance

A silica fiber is linear at low data rates and low power levels. Linearity means that, if two signals are present, they will be treated the same by the system or component and will not interact. A silica fiber will have a nonlinear response at high data rates and high power, however. Some of the effects on fiber performance can be useful and others detrimental. Nonlinearity is caused by two basic phenomena inherent in the fiber: scattering and refractive index variations. Scattering is caused by interactions between an optical signal and molecular or acoustic vibrations within the silica. Refractive index variations occur because the refractive index can depend on changes in light intensity.

Scattering occurs as *stimulated Raman scattering* (SRS) and *stimulated Brillouin scattering* (SBS). Refractive index variations cause *cross-phase modulation* (XPM), *self-phase modulation* (SPM), and *four-wave mixing* (FWM),). The SRS, SBS, and FWM processes depend on optical signal intensity and result in gains or losses in power within the fiber. Unwanted crosstalk in WDM systems is one result. On the positive side, amplifiers are now available for the S band (1480 to 1520 nm) and L band (1570 to 1620 nm) that rely on Raman scattering. These now offer significant expansion of WDM system capability and the efficient use of fiber already in place. Both XPM and SPM affect the phase of a signal, which can cause chirping (increased linewidth) in high data rate systems.

SRS causes an interaction between the optical signal and the molecules in silica. SRS transfers optical power from shorter wavelengths to longer wavelengths. Independent pump energy is required, but the phenomenon is nonresonant. The gain increases with wavelength distance from the pump wavelength. In the 1550 nm region, for example, the gain peaks at 125 nm above the pump wavelength. Gain is absent beyond 1675 nm. SRS is an important factor in determining the performance of long amplified, WDM systems. The shortest wavelength of the WDM group will have significant power transferred to higher wavelengths. To assure a performance power penalty <1 dB in a DWDM system from SRS, the maximum power per channel in the fiber should be kept to less than a few dBm.

The SRS phenomenon is used in Raman amplifiers recently developed to provide gain in the L, C, and S bands (1570 to 1620,1525 to 1565 and 1440 to 1520nm respectively). EDFA amplifiers have been used for some time to provide gain in the C band. Now EDFA amplifiers are combined with Raman amplifiers in a hybrid configuration. Raman amplifiers are implemented in distributed or lumped form. A distributed amplifier, for example, requires a pump power of 1 watt at 1500 nm for a conventional SM silica fiber and will provide about 30 dB gain across the C band. Gain equalization is required since the amount of transferred energy varies across the WDM channels. A combination of a Raman distributed amplifier with a section of dispersion compensating fiber is available commercially. A lumped Raman amplifier has the required fiber spooled inside the amplifier housing.

Stimulated brillouin scattering (SBS) is caused by lightwaves interacting with acoustic waves in a SM fiber. Since acoustic vibrations are random occurrences in the fiber, the long-term system degradations caused by SBS are difficult to quantify. The light is scattered backward from the direction of transmission. The backscatter extracts gain from the main signal. It is confined to a small bandwidth offset lower in wavelength from the forward signal; hence, SBS is of concern mainly in WDM systems. Since isolators are used to protect amplifiers from echoes, SBS effects are confined to the span where vibrations are occurring. The system power penalty from SBS becomes significant when the scattered wave power is comparable to the forward power. Once the total input power to the fiber reaches about +10 dBm in silica, any further power increases will transfer completely to the backscattered signal. Ironically, the threshold power is higher if the optical source bandwidth is wider. SBS can be controlled by using some or all of the following: lower power levels, wider linewidth source, and purposefully applying a low-frequency dither to the source.

Self-phase modulation (SPM) results when the propagating optical carrier induces phase modulation on itself. This is caused by the fiber's index of refraction (η) having a (relatively weak) dependence on the light intensity.

The nonlinearity, which is called the Kerr effect, is important in achieving Soliton transmission (Section 7.2.4.8). Pulses are either spread or contracted in time, depending on the central wavelength's value, the chromatic dispersion, and the pulse's shape and energy.

With WDM systems, *cross phase modulation* (XPM) occurs because of the same Kerr effect. SPM and XPM are more pronounced when the pulse intensity and bit rate are high. At 10 and 40 Gb/s, XPM can add to the effects of polarization mode dispersion (PMD). XPM has been found to induce rapid changes in the state of polarization in an affected high data rate channel. PMD equalizers are not able to track this additional cause of pulse spreading. The use of lower powers and/or forward error correction can alleviate the problem.

Four-wave mixing (FWM) leads to third-order intermodulation distortion that can be significant, particularly in dense WDM systems (DWDM). FWM occurs when three wavelengths mix and generate a fourth, interfering wavelength. The new wavelength can fall back on one of the original three, thus causing severe crosstalk. The severity of FWM depends on the relative coherence of the primary waves. If they are located at or near the zero chromatic dispersion wavelength, the effect of FWM can be significant because their relative coherence is maintained. Four-wave mixing interference is more severe when the group velocities of the primary wavelengths are close. It is lessened if the group velocity differences are higher and the wavelength spacing wider. This is the reason *nonzero dispersion shifted fiber* (NZDSF) is used for 1550 nm high data rate systems. With NZDSF the attenuation is low and there is a small amount of chromatic dispersion in the transmission bands of primary interest, C and L. The small values of chromatic dispersion result in acceptably low pulse spreading but are sufficient to minimize the effects of FWM.

On the positive side, FWM is one of the techniques used to accomplish wavelength conversion at terminal locations. Wavelength converters translate information from an incoming wavelength to a new wavelength without using O/E and E/O conversions. These conversions are very useful in optical networks. The FWM takes place in a nonlinear medium, active or passive, through interaction between optical waves. FWM converters can be used for both analog and digital signals because the phase and frequency of the generated wave are a linear combination of the interacting signals. Further, FWM converters allow multiple signals to be processed at the same time.

7.2.4.6 *Crosstalk*

Crosstalk in optical systems causes problems similar to that faced by radio systems with closely spaced carriers. If the interference comes from another channel in a WDM system, it is *interchannel crosstalk* (ITXT). If the interference comes from the desired signal, it is *intrachannel crosstalk* (IAXT). ITXT can originate in wavelength multiplexers/demultiplexers, optical switches, optical filters, amplifiers, and fibers. IAXT can occur in some types of optical switches. ITXT is less troublesome to a signal than IAXT if the interfering levels are equal. The larger the number of WDM channels, the greater the potential for crosstalk problems. With analog systems, the crosstalk adds to the system noise. The high levels of S/N needed for analog signals make dense WDM systems impractical: the crosstalk would be too high. With digital systems, if ITXT rejection is > 30 dB there will be less than a 1dB added power penalty in a system with eight channels. The IAXT would have to be < –10 dB down. The use of multiple devices in the transmission paths of optical networks leads to an increased potential for crosstalk. A simple end-to-end system that contains only a few devices will probably not have significant crosstalk.

7.2.4.7 *Solitons*

The Kerr effect, introduced in Section 7.2.4 relative to XPM and SPM, occurs when fast changes in light intensity modulate the refractive index of a silica fiber. The effect is very small and not noticeable in analog systems or digital systems modulated below 10 Gb/s. SPM occurs when the energy in the leading and trailing edges of the pulse change rapidly. The refractive index at the leading edge of a pulse sees a positive $d\eta/dt$ while the trailing edge sees a negative $d\eta/dt$. Because of the Kerr effect, the leading edge of the carrier pulse undergoes a positive phase shift and the trailing edge a negative phase shift. The positive phase shift increases the effective wavelength on the leading pulse edge. The negative phase shift decreases the effective wavelength on the trailing pulse edge. If the chromatic dispersion is negative (M positive in Figures 3-4 and 3-5), the leading edge travels faster and disperses away from the center of the pulse. The trailing edge moves slower and also disperses away from the center. This is a *chirp,* or spectral broadening, which worsens the pulse broadening caused by chromatic dispersion. When chromatic dispersion is positive (M negative), the leading edge travels slower and moves toward the center of the pulse. The trailing edge travels faster and also moves toward the center. The resulting pulse narrowing partly compensates for the chromatic dispersion and forms the basis for soliton transmission. Figure 7-6 shows these relationships for two pulses, one with its central wavelength above the dispersion null and the other with its central wavelength below. When dispersion shifted fiber is used for high bit rate systems, M is low but the refractive index's time rate-of-change dependency is high. This results in XPM and SPM becoming more pronounced.

Solitons are very short pulses with specific widths and peak energy. They use the RZ line code with a low duty cycle. Solitons use the small fiber

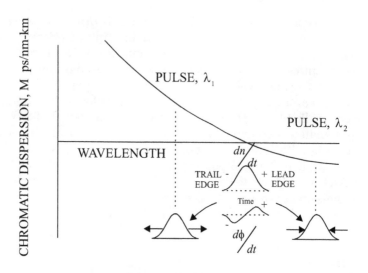

Figure 7-6 *Kerr effect (XPM, SPM, solitons).*

intensity dependency described by the Kerr effect and are transmitted at wavelengths where the chromatic dispersion M parameter is negative, i.e., at wavelengths above the dispersion null. An important parameter, the product of width and pulse energy, is held constant by the interaction between the fiber's dispersion and nonlinearity. The soliton then can propagate without spreading. There are two basic types of solitons: fundamental or first order, and higher order. A fundamental soliton does not change shape as it propagates. A higher order soliton has periodic fluctuations, or pulsations, in its shape. Solitons lose energy as they propagate because of the fiber's attenuation, so they must have periodic amplification. It has been found that distributed rather than lumped amplification is best for solitons.

Solitons are a delicate form of transmission. If the energy is too low, the soliton will not form and the pulse will spread because of chromatic dispersion. If the energy is too high or the pulse is the wrong shape, a higher order soliton will propagate. Extra energy beyond that needed to form and sustain the soliton generates a spurious, interfering, companion soliton pulse.

7.3   Link Budgets

Initially, fiber optic systems were only used in point-to-point, or *link* connections. The SONET/SDM lengths listed in Table 7-2 are link lengths.

A fiber optic link has the basic elements shown in Figures 1-2 and 7-1: source, fiber, and receiver. First and second-generation systems consisted only of these point-to-point link connections. The first generation systems used multimode fiber in the 800 nm window; the second generation used standard single-mode fiber in the 1300 nm window. Later, 800/1300 nm WDM and optical amplifiers were added, but the links still remained point-to-point and uncomplicated from a networking viewpoint. The link design considered these basic requirements:

• Baseband signal type, analog or digital
• Required performance, S/N for analog or BER for digital
• Top frequency or data rate
• Transmission distance

Sections 7.3.1 and 7.3.2 analyze two specific link designs. These illustrate many of the tradeoffs available in the design of optical fiber systems. Link power and rise time budgets are calculated for both systems. They ensure a desired system performance for the specific baseband signal. The system design variables are the operating characteristics of the three main system components: source, fiber, and receiver. Table 7-3 lists the variety of choices available for the parameters that affect system performance. Not all of these need be included directly in link analyses. Some parameter choices evolve from other choices. For example, choosing a SM fiber will always result in a NA in the region of 0.2.

Table 7-3 *System design parameters.*

SOURCE- LED or Laser	PARAMETER
	wavelength
	linewidth
	power out
	radiation pattern
	long. mode structure
	speed of response
FIBER- MM or SM	
	bandwidth/dispersion
	polarization mode dispersion (PMD)
	attenuation
	core size/profile
	numerical aperture
PHOTODETECTOR- PIN or APD	
	wavelength
	sensitivity
	responsivity
	speed of response
	load resistance

Two separate analyses are needed to obtain a satisfactory picture of an optical link's planned performance. One, the link power budget, examines the power launched/received and accounts for the losses of intermediate components. The other, the link rise time budget, deals with component rise times. A successful link design will find combinations of source, fiber, and receiver that satisfy both budgets. In all likelihood, one budget will be met more closely than the other by a proposed system design. The end result of the two analyses will be a fiber length that meets the initial objectives and also allows the received signal power and anticipated noise/distortion to fit within the two budgets.

In this chapter, two different optical fiber systems are adopted to illustrate the range of system applications: a low cost, analog, video distribution system for intra-campus security use and a 2.5 Gb/s long haul terrestrial system. The latter system uses WDM to expand capacity and optical amplifiers to obtain a long distance. A long haul system was chosen whose length or reach, 1000 km, is at the upper limit of links used in the current national network. This leads to the use of an APD receiver to gain maximum sensitivity. Links between regenerators currently average 600–800 km in the national network. On most links a lower cost, less sensitive PIN-FET receiver would be adequate. The longer system was chosen to bring as many factors as possible into the example.

Both the power and rise time calculations have link length as a governing parameter. Not all noise and distortion contributions are length dependent, however. These include non-zero extinction ratio, modal noise, laser intensity noise, reflection noise, and chirp. Part of the system power margin has to be set aside to protect against their negative effects on performance. The system power margin is obtained by purposefully designing the S/N ratio at the receiver to be higher than the minimum required to overcome receiver noise alone. The extra degree of protection shields against these difficult-to-quantify degradations, as well as the effects of component aging.

7.3.1 *Analog Video System*

Optical fiber systems tend to find use in two different applications: short distance local areas using 800nm/MM and longer distances using 1300 to 1500 nm/SM. This video analog system is in the first category. It is designed to transmit remote video camera signals to a central monitoring location. The maximum distance is 1.5 km; the baseband signal is NTSC color video. The total signal bandwidth is 6 MHz, but the noise bandwidth for the video signal (which conveys only luminance) is 4 MHz. The sound and color subcarriers carried above the video signal have separate formats for modulation/demodulation. Their performance is not analyzed here. A 50 dB S/N ratio is required for the luminance signal. Low cost is very important for this system. As a result, the following components have been chosen:

- Surface emitting LED source at 850 nm
- GMM fiber
- Silicon PIN diode, FET preamp

7.3.1.1 *Analog Link Power Budget*

Table 7-4 gives the link power budget for this analog system. A connector at each end is needed, one on the source pigtail and the other on the input to the photodetector. Because this is probably an underground installation, the length of 1.5 km is assumed to require an intermediate splice. After detection and amplification, the video signal goes directly to a dedicated monitor.

Several assumptions have been used for the PIN-FET receiver. The parameters and values are taken from Figure 5-8 and Table 5-2:

- I_D = dark current = $10^{-9} A$
- $C_T = C_A + C_D = 2$ pF +3 pF = 5 pF
- $R_T = R_L = 5000$ ohms
- F = preamp noise figure = 3 dB = 2

The load resistor value of 5 k results in a 3 dB roll-off frequency above the highest signal frequency of 6 MHz. From Equation 5.12,

$$f_{3-dB} = \frac{1}{2\pi R_T C_T} = \frac{1}{6.28 \cdot 5(10^3) \cdot 5(10^{-12})} = 6.3 \; MHz$$

This value of load resistor satisfies the baseband signal bandwidth requirement of 6 MHz. Further, the need for equalization is avoided because Equation 5.14 is satisfied:

$$R_T < \frac{1}{2\pi C_T B} = \frac{1}{6.28 \cdot 5(10^{-12}) \cdot 6(10^6)} = 5.3 \; k\Omega$$

Table 7-4 *Analog link power.*

	Loss, dB	System results
LED power coupled to fiber		-10 dBm (0.1 mW)
source connector loss	0.5	
fiber loss (8dB/km x 1.5 km)	12	
splice loss	0.1	
detector connector loss	0.5	
detector reflection loss	0.2	
Total Loss		13.3 dB
Power into PIN		-23.3 dBm (0.0047 mW)
S/N with rcvd. power		50 dB
Power required for S/N=50 dB		-30.6 dBm
Margin (30.6-23.3)		7.3 dB

A S/N of 50 dB for the received signal is calculated from Equation 5.16 as follows:

$$S \backslash N = \frac{\frac{m^2}{2} I_{out}^2 R_T}{2q(I_{out} + I_D) BR_T + 4kTBF} = 10^5 = 50 \text{ dB} \qquad 7.11$$

where I_{out} = rms diode current = RP_{in} = (0.6) A/w (4.7 µW) = 2.82 $10^{-6} A$, m = index of AM = 0.75, B = noise bandwidth = 4 MHz. An index of modulation of 75% is customarily used. The S/N and input power will not track dB for dB. If the length is increased to 2 km, the power, P_{in}, drops by 4 dB. The 7.3 dB margin is reduced by more than 4 dB because the power change is squared in the numerator of Equation 7.11 but is not squared in the denominator. This means that care must be taken when using a margin calculated early in the design process to compensate for component performance. The S/N is dominated by the thermal noise. This will normally be the case with PIN-FET analog receivers. The AM modulation index appears because the signal is continuous and requires the use of the rms value in noise calculations. In the next section, a digital noise calculation will use only the average signal power value. This assumes a constant signal value throughout the period when a one is sent, and no signal when a zero is sent.

As with all communications engineering, many trade-offs are possible in designing a system. This initial design provides a relatively small margin to protect against design errors, component drift, components out-of-spec, or environmentally caused performance changes. There are several ways to increase the margin. A 100 percent index of modulation would increase the signal power but could lead to distortions caused by nonlinearity in the LED source. In the opposite direction, reducing the index to 50 percent to improve potential distortion problems would decrease the S/N by $(0.5/0.75)^2$, thus decreasing the margin by 6.5 dB. A laser source would improve linearity but might add cost to the system. Coupled power would be increased by about 10 dB with a laser, thus yielding more margin protection. Vertical cavity surface emitting lasers (VCSEL, Chapter 8) are low in cost but an LED would have better linearity.

With the PIN-FET low input resistance receiver, the S/N is controlled by thermal noise. A larger value of R_T would make the shot and thermal noise more equal as well as increase the S/N. The drawback would be the need for equalization in the receiver to compensate for roll-off. Moreover, a larger load resistor would decrease the dynamic range. If a photodetector power source (V_{PS}) of 5 volts is assumed, the 5 kΩ load resistor with the high input impedance preamp provides sufficient dynamic range. The peak power at the receiver will be about −22 dBm (6.3 µW). Using the characteristic curves in Figure 5-6 as an example of photodetector response, saturation will not be a problem. If the load resistor is increased to 50 kΩ and the need for equalization accepted, the following would result: S/N would increase, thus giving better margin and the saturation level would drop, causing possible distortion in the receiver. An APD could be used in place of the PIN to gain better photodetector sensitivity and S/N without sacrificing range. However, the extra costs would be significant. In summary, to gain more system margin, it is possible to use some or all of the following: a higher resistor value, equalization, a laser, or perhaps even slightly better fiber with lower attenuation.

An alternative in providing short distance video transmission is an all electrical coaxial cable system. Air-dielectric coaxes have very good bandwidth and could be used for the 1.5 km distance. Attenuation per kilometer would be higher, but for very short systems the two approaches would result in similar performance. A bandwidth of 6 MHz presents no problem for the baseband electronics for either system. Not all trade-offs are technical, however. The air-dielectric coax would have to be pressurized to keep out moisture. The fiber cable does not need this protection. The metallic coax system is subject to EMI and lightning damage; the fiber is not if an all-dielectric cable is chosen. The fiber cable would be smaller, less bulky and probably less expensive. Taking all factors into account, the fiber system seems a better choice if cost is the primary consideration.

7.3.1.2 *Analog Link Bandwidth Budget*

Chapter 3 discussed the basic concepts behind time-of-arrival distortions at the optical level. The electrical baseband parts of the system do not affect this distortion. The fiber's delay distortion caused by dispersion requires the development of a rise time value to complete the total system transmission analysis. Assuming the system and its components are modeled as having first order responses, Equation 3.4, restated below, relates available bandwidth and the resulting rise time:

$$f_{3dB\ electrical} = \frac{0.35}{t_r}$$

where $t_r = \Delta\tau = 10 - 90\ \%$ system or component rise time and $\Delta\tau$ = delay spread. Because the rise times of each component act independently, the total system rise time can be obtained through a root-sum-square calculation.

$$t_{r\text{-}system} = (t_{r\text{-}source}^2 + t_{r\text{-}fiber}^2 + t_{r\text{-}receiver}^2)^{1/2} \qquad 7.12$$

If the fiber has both modal and chromatic dispersion, they are treated as independent contributors. Usually, one or the other will dominate. A tabulation of the rise times for this simple analog system is given in

Table 7-5 *Analog link bandwidth.*

	Component rise time, ns	System rise time, ns
LED rise time	25	
fiber, modal	0.046	
fiber, chromatic	9	
photodetector/receiver	33	
Total Rise Time, T	42.4	
System requirement (0.35/6MHz)		58
Margin (Requirement-T)		15.6

Table 7-5. The surface emitting LED is assumed to have a rise time of 25 ns, and a FWHM optical bandwidth of 50 MHz. The GMM fiber has the characteristics from Chapter 3 (η_1 = 1.5, *NA* = 0.173, Δ = 0.007). Applying equation 3.9, the modal distortion is 0.046 ns. Using a broad linewidth LED at 800 nm causes chromatic dispersion to become more important than modal dispersion. Using equations 5.12, 13, the photodetector/receiver's rise time of 33 ms. is based on the load resistor (5k) and the diode's (3 pF) capacitance in parallel ($t_r = 0.35/C^{1/2}\pi R_T C_T$). Transit time for this device would be on the order of 0.75 ns (Table 5-2), thus the circuit rise time dominates. As a rule of thumb, transit time effects can be ignored if they are less than one fourth the circuit rise time. The total system rise time becomes:

$$t_{r\text{-system}} = (25^2 + 9^2 + 33^2)^{1/2} \ ns = 42.4ns$$

As with the power budget, some observations need to be made about possible tradeoffs. The pulse spread caused by chromatic dispersion alone was calculated based on a value of 120 ps/(nm-km) at 800 nm (Figure 3-4). Using a step index fiber instead of the GMM fiber would increase modal dispersion by a factor of about 1000, thus making it impossible to meet the system rise time objective. The spectral width of the (LED) source could be narrowed significantly by using a laser. This would give more flexibility to the receiver design but add cost to the system. At the receiver, the PIN diode has good bandwidth performance. A 5 pF diode would increase rise time significantly and again make it difficult to meet the system rise time objective, all other factors remaining constant. One of the trade-offs available involves the load resistor (R_T). If the load resistor is decreased by a factor of 2, the photodetector rise time improves by the same factor but the S/N would be worsened by about 3 dB.

7.3.2   Digital Link

Optical fiber installations exploded in the 1990s. A synergistic relationship developed between the perceived need for cheap, reliable, error-free digital communications and the constantly improving broadband capability of single mode fiber systems. The digital system model used here reflects this marriage by using a third generation 1550 nm, point-to-point, four channel WDM system design. The model uses optical bandwidth more conservatively than the most advanced dense WDM (DWDM) systems. The four channels are spaced 1.6 nm apart (200 GHz at 1500nm). A truly dense WDM system would have channels packed as close as 25 GHz (0.2 nm at 1550nm). Achieving this dense a channel packing requires higher cost components, however. One of the objectives for this model system is to have the costs under reasonable control.

Systems in the third window have seen constant performance improvements since first being introduced in the early 1990s. The loss in silica fiber is the smallest in the range of 1550 nm (Chapter 3). The initial 1550 nm systems used previously installed 1300 nm standard SM fiber. This fiber has a chromatic dispersion zero at 1310 nm, and about 17 ps/nm-km at 1550 nm (Chapter 3). The chromatic dispersion is large enough to limit system length at 1550 nm because of excessive pulse spreading. As a result, a number of methods to mitigate pulse spreading due to chromatic dispersion in the third window have been developed: dispersion shifted fiber, non-zero dispersion shifted fiber (NZDSF), dispersion compensation devices, dispersion equalization, single-longitudinal mode lasers, external modulators. Finally, optical amplifiers are now available to better manage fiber and component loss, extend system lengths, and lower costs. The system model adopted here uses NZDSF and EDFA optical amplifiers. The use of amplifiers leads to the system design being dispersion, not power limited.

First and second generation digital systems used periodically distributed electrical regenerators to achieve long transmission distances. Electrical regenerators perform a number of tasks: O-E conversion, extraction of a timing signal, deciding whether a 1 or 0 is present, pulse regeneration, and finally E-O conversion. The third generation system modeled here uses Erbium doped fiber amplifiers (EDFA) instead of regenerators. In this design, the EDFAs are placed uniformly through the total system length of 1000 km at fiber span lengths of 100 kilometers. Studies have shown that if EDFAs are used at intermediate locations, they are best placed with uniform spacing.

A schematic of one of the four WDM channels is shown in Figure 7-7. The optical source is a directly modulated SLM (DFB) laser with a linewidth of 0.1 nm. The laser package includes an isolator and has a pigtail that appears on a patch panel. Patch panels, which provide flexibility for service routing, maintenance, and reassignments, are assumed on each side of the WDM devices at both terminal locations. The WDM filters at the terminals are pigtailed and appear on the panels with a connector. Amplifier locations have similar patch panels plus climate control and uninterruptible power.

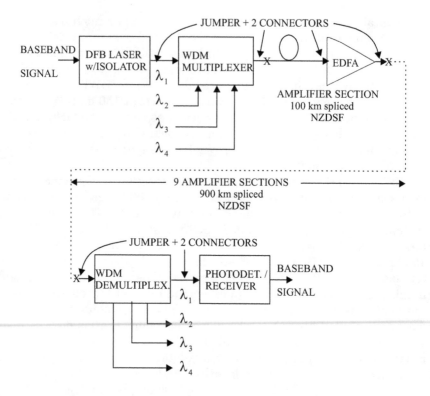

Figure 7-7 *Digital system model.*

When used in metropolitan optical networks, this requirement is easily met because locations have numerous and closely spaced *optical add / drops* (OAD). Each amplifier package is assumed to have an optical isolator at the output to limit reflections. In addition, an optical filter is assumed that rejects out-of-band ASE noise and compensates for the nonuniform gain shape of the EDFA. The receiver front end is comprised of an APD and transimpedance amplifier.

This channel is assumed to be one of four in a C band WDM system. Each of the four channels carries a 2.5 Gb/s NRZ baseband signal. The optical filters are assumed to be 0.6 nm wide at the 0.5 dB optical wavelengths. The filters for each channel will give 25 dB of isolation at the next adjacent channel wavelength, thus limiting crosstalk. This calls for a filter with a figure of merit of about 2 (Chapter 6). Transmitted powers are kept relatively low so that crosstalk caused by fiber nonlinearities will be absent. If the channels were more closely spaced, the laser source would probably require expensive temperature control to limit center wavelength drifting. The

WDM devices combine the filtering and multiplexing/demultiplexing requirements and are assumed to be micro-optic grating devices, as opposed to thin film devices. Micro-optic gratings have lower insertion losses because all channels are handled in parallel.

The four WDM channels are each point-to-point, but their signals have to traverse shared WDM optical devices and amplifiers. The gain of each amplifier offsets the fiber and component losses in the individual spans. Since an EDFA has unequal gain across the C band, gain equalization is assumed at each amplifier location. Each of the four channels has the same design. The fiber is single mode, nonzero dispersion shifted fiber (NZDSF). The choice of NZDSF is a compromise that minimizes 4-wave mixing distortion, XPM and SPM, and the effects of chromatic dispersion.

As with the analog system, two basic areas of performance of the digital system have to be analyzed: power losses and gains, and bandwidth. The former is analyzed through the development of a power budget. The latter is analyzed through the development of a rise time budget. If the designer wants to examine the tradeoff between distance and bit rate, relationships like those plotted in Figure 7-8 need to be developed. The calculations that generated the Figure 7-8 plots used the component characteristics of the system adopted here.

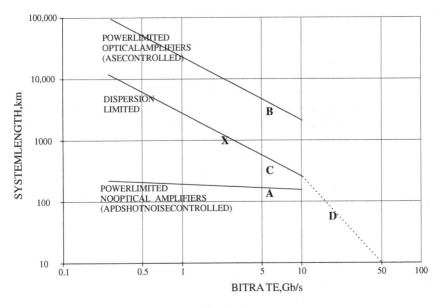

Figure 7-8 *Digital system design limitations.*

Note that there are four lines, labeled A to D, in Figure 7-8. These plots give general solutions. The specific digital link with the components listed earlier has the operating point X. Line A gives the power-limited, nonamplified relationship between total distance and bit rate for a single span. Line A is generated by solving Equation 7.15 for the required receiver sensitivity as a function of bit rate. The system performance objective used was an S/N (average values) of 16.4 dB, which will give a BER of 10^{-11}. Line A is shown for reference only. It plots the (traditional) view that occurs when receiver noise, not optical amplifier noise, limits the power budget and resulting system length. The slope is a function of bit rate because the receiver shot noise contribution, hence the S/N, depends on bandwidth. The top line (B), gives the power-limited, amplified relationship between possible system lengths and bit rate. The long distances in curve B are achieved through the use of amplifiers in this system. The amplifiers make fiber attenuation and component losses unimportant in determining the power margin, and the amplifier's ASE noise now becomes an important consideration in the design process. The slope for B is dictated by the dependency of amplifier noise on bandwidth and attenuation. ASE controls these results. Lines C and D give the dispersion-limited relationships. The total length of all the amplified spans, which is 1000 km, causes the pulse spreading. Fiber dispersion is not compensated in this system design. The slope for C is caused by the chromatic dispersion-controlled spread. Recall that the effects of spread are dependent on the inverse of the bit rate. The governing relationship is

$$L = \frac{0.5}{\Delta\lambda \cdot D_{CHR} \cdot B_{NRZ}} \text{ km} \qquad 7.13$$

where D_{CHR} = chromatic dispersion in ps/(nm-km) and $\Delta\lambda$ = laser linewidth in nm. For this system, the laser has an assumed linewidth of 0.1 nm and the chromatic dispersion is 2 ps/(nm-km). The right-hand line (D) represents a speculative design that might exist for (future) 10 and 40 Gb/s systems. These systems will be dispersion limited and use monochromatic lasers (linewidth < 50 MHz) with external modulators. As a result, the transmitted optical spectrum will be equal to the modulation bit rate, not the $\Delta\lambda$ of the laser. Achievable system lengths then become dependent on the square of the transmitted bit rate. The governing relationship is:

$$L = \frac{c}{2 \cdot D_{CHR} \cdot \lambda^2 \cdot B_{NRZ}^2} \text{ km} \qquad 7.14$$

At 10 Gb/s the allowable length based on equation 7.14 is about 250 km. For the fixed value of chromatic dispersion used in this example, 2 ps/nm-km, system lengths drop by a factor of 16 when the bit rate goes from

10 to 40 Gb/s. As a result, dispersion control or compensation will be required for these systems (Chapter 8).

A specific system design is generally acceptable if it is on or under the applicable envelopes A-D. The model used in this section lies at point X. The design just meets the dispersion objective but has a significant power margin thanks to the use of optical amplifiers. This advantage can be seen by comparing point X to line B. The extra signal power (S) is available to protect against added noises. In the eye diagram (Figure 7-5), these noises would appear primarily on the vertical axis at the decision point.

On Figure 7-3, a measured BER curve for the system would be curve D, displaced significantly from curve B. An average signal level of −35.4 dBm at the APD gives 10^{-11} BER when there are no added noises beyond those of the receiver. Curve D illustrates what would happen with the added ASE noise from the amplifiers. The 10^{-11} BER would be achieved at a 10 dB higher average received signal. This would be the penalty associated with the live amplifiers.

7.3.2.1 *Digital Link Power Budget*

A power budget analysis of the model digital system is listed in Table 7-6. The output power of the laser package and the amplifiers is kept low so that nonlinear effects in the fiber will be minimized. Sometimes optical attenuators have to be used to control signal levels. The assumed loss per kilometer of 0.25 dB for the SM fiber includes splice loss. The gain of the intermediate amplifiers (30 dB) is in the upper range of gains available with EDFAs. The input power to the amplifiers is on the low side of possible input levels for EDFAs. The last amplified span serves as an optical preamplifier at the receiver. A separate optical preamp would not improve the receiver sensitivity because the receiver performance is already shot noise limited. The 20.4 dB margin (after 10 dB ASE penalty) is used to maintain the required system BER in the presence of all noises/distortions other than receiver and ASE noise. These other noises are difficult to quantify, which means the system design should have an adequate power margin to protect service quality. Assuming a 50% eye closure due to ISI, about 7.2 dB of margin remains for unanticipated degradations.

Figure 7-3 gave plots of the bit-error-rate vs. receiver S/N. The system design discussed here must meet the SONET error rate objective of 10^{-11} which requires a transceiver S/N of 17.4 dB when the ISI penalty of 2 dB is included. The ISI penalty accounts for the effects of dispersion, mode partitioning, chirp, RIN, and jitter.

By applying Equation 5.18, the S/N objective can be used to determine the required receiver sensitivity. The InGaAs-APD photodetector/receiver parameters used are listed in Table 5-2:

Table 7-6 *Digital power budget.*

		Loss or (Gain) - dB	Power (dBm
Laser package out average power (A)			+5
	Connector	0.5	
	Jumper	1.0	
	Connector	0.5	
	WDM device	1.0	
	Connector	0.5	
	Jumper	1.0	
	Connector	0.5	
Average Power into fiber (B)			0
	100 km fiber @0.25 dB/km	25	
Average Power into amplifier (C)			-25
	Jump. +2 Conn.	2	
	Amplifier Gain	(30)	
	Filter-Equaliz.	1	
	Jump. +2 conn.	2	
Average power out amplifier (D)			0
	Connector	0.5	
	Jumper	1	
	Connector	0.5	
	WDM device	1	
	Connector	0.5	
	Jumper	1	
	Connector	0.5	
Average power into Receiver (E)			-5
APD sensitivity @ 2.5 GHz			-35.4
Beginning Margin (35.4-5)			30.4 dB
ASE penalty		10	
50 % eye closure requirement		13.2	
Remaining margin			7.2 dB

$$S/N = \frac{(RP_{in})^2 \, M^2}{2q((RP_{in}) + I_D)M^2 M^x \, B + \dfrac{4kTBF}{R_T}} \qquad 7.15$$

where

$\dfrac{S}{N} = 10^{1.74} = 55, M = 30, x = 0.5, R_T = \text{high} \approx 10^5 \, \Omega, T = 300°K$

$B = 2.5(10^9), R = \text{responsivity} = 1, F = \text{noise figure} = 2$

Solving for the input power that gives the required BER and corresponding S/N,

$$P_{in} = 0.241(10^{-6}) \, W = 0.24(10^{-3}) \, mW = -36.2 \, dBm$$

The detector is shot noise limited because the photodetector current in the denominator gives the dominant noise term. In the absence of any other noise factors, the −36.2 dBm input power is the minimum level for this receiver that will result in satisfactory (SONET) performance. With the ASE noise the signal cannot drop below −26.2 dBm. Finally, to always meet the 40% eye closure, the signal has to be above −13 dBm.

Note that laser noise (RIN) has not been included in determining the back-to-back transceiver performance (curve B). RIN is always low and can be ignored as a direct contributor to receiver noise in digital systems. It can be useful, however, in a different approach to calculating the effect of ASE noise from cascaded amplifiers. An analysis (not included) was made that treated the cascaded amplifiers as one noise producer that would yield an overall equivalent system noise figure. For this calculation, the input S/N to the line was the ratio of laser output power to RIN. The gain and noise figure used for each amplifier were the same as those used in Equation 7.7. The results were comparable to those based on the ASE calculations in Section 7.2.4.2.

The optical filter/equalizer at each amplifier provides two functions: it filters noise outside the channel bandwidth and equalizes the amplifier gains between channels. A broadband EDFA has an unequal gain shape. If not corrected, the receivers would have large power differences between channels at the end terminal location. This design assumes the optical filter and equalizer contribute a combined insertion loss of 1 dB. Also, the amplifiers are assumed to be rack mounted and accessed through patch panels. With the gain set at 30 dB, the allowable length between amplifiers is 100 km ((30-5)/0.25). The output power at each intermediate amplifier is 0 dBm. This is also the power launched into the fiber at the transmitting location. Output powers can range up to 20 dBm for EDFAs, but levels that high in a DWDM system configuration would cause concern about the effects of fiber nonlinearities (Section 7.2.4.6). The power sum from all channels in a WDM system has to be considered. In this system of four channels, each with 0 dBm amplifier outputs, the average total power into the fiber is +6 dBm. Peak powers can be higher, so the approach used here is to keep the total average power below +10 dBm.

A long distance, digital, high data rate WDM system, such as that modeled here, is more vulnerable to additional noise than a short distance analog system. Curve C in Figure 7-3 shows a sample receiver BER performance when additional noise and degradations are present that follow the signal level dB for dB. As the S/N rises, these noises become controlling. Curve D in Figure 7-3 shows the effects of ASE noise alone in the model channel. At a 10^{-11} BER, the required increase in signal needed to offset the ASE was 10 dB (Section 7.2.4.2). This uses a large portion of the available 30.4 dB margin. From the eye analysis (section 7.2.4), 13.2 dB of the margin is needed to offset ISI. Other factors can further erode the 7.2 dB of remaining margin: mode partition noise (MPN), extinction ratio noise, chirp, crosstalk, reflection noise, modal noise, component aging and drift. Exact quantification of the power penalties for most of these possible noise contributions is not possible. The correctness of the system design has to be verified later through measurements. A conservative approach should be adopted during the design phase.

7.3.2.2 *Digital Link Rise Time Analysis*

This section analyzes rise times and quantifies the effects of ISI. In the eye diagram of Figure 7-5, ISI and jitter cause displacements on the horizontal and vertical axes. The calculations performed later with rise times show that each of the four channels will be dispersion limited. With a dispersion limited channel, there is extra S/N margin in the power budget. This means the signal at the receiver is high enough to protect against added noises from RIN, mode partitioning, etc. Some factors can cause additional S/N deterioration but are difficult to quantify. For example, PMD was calculated at 0.016 ns for this system (section 7-2-4-4). This is not large enough to be a factor in the rise time calculations. However, PMD is a random phenomenon. Environmental conditions can change the instantaneous effect of PMD. The power budget margin of 7.2 dB will provide protection against possible increased ISI and jitter caused by the environment.

The rise time/bandwidth assumptions and analytical results are listed in Table 7-7. Recall from equation 3.5 that, for NRZ pulse, the rise time equals the delay. The system objective is derived according to Equation 7.3.

$$t_s = \text{System rise time objective} = \frac{0.707}{\text{bit rate}} = \frac{0.707}{2.5(10)^9} = 0.28 \ ns \qquad 7.16$$

From equation 3.16, the fiber chromatic dispersion causes a pulse spread of:

$$\Delta\tau_{chr} = -\Delta\lambda \cdot L \cdot D = 0.2ns \qquad 7.17$$

Table 7-7 *Digital link bandwidth.*

	Component rise time, ns	System rise time, ns
Laser rise time	0.025	
fiber, chromatic	0.2	
fiber, PMD	0.016	
photodetector/receiver	0.15	
Total Rise Time, T	0.252	
System requirement		0.28
Margin (Requirement-T)		0.03

where $\Delta\lambda$ = laser linewidth = 0.1 nm, L = total fiber length = 100 km, D = chromatic dispersion at 1550 nm = 2 ps/nm x km

From Table 4-3, the rise time of the laser source is assumed to be 25 ps. The rise time of the InGaAs APD alone, if controlled by transit time, would be on the order of 0.15 ns (Table 5-2). By using a transimpedance pream-plifier, the circuit bandwidth is enhanced by the gain in the op-amp config-uration. Using Equation 5.12, a C_T = 1 ps and assuming a gain of 150, the bandwidth becomes 2.5 GHz. The circuit limited rise time then would be 0.14 ns (Equation 5.13). This is about the same as the APD transit time response alone. Since the preamp rise time can be improved in redesign, if required, the assumption used here is that the receiver rise time (t_R)is dic-tated by the photodetector transit time.

The rise time total from the three basic components then becomes:

$$t_{SYSTEM} = \sqrt{(t_{LASER})^2 + (t_{FIBER})^2 + (t_{RECEIVER})^2}$$

$$t_{SYSTEM} = \sqrt{0.025^2 + 0.2^2 + 0.15^2} \; ns = 0.252 ns$$

7.18

The rise time is close to the objective of 0.28 ns, which makes the channel design dispersion limited. This is illustrated by the location of point X in Figure 7-8.

A brief review and analysis of the methodologies used to translate time spread (rise time) into system performance seems appropriate at this junc-ture. An intuitive approach was used in this text to obtain the fiber's band-width. It examined the time spread between transmitted wavelengths by looking at two baseband frequencies that combined in the receiver 180° out of phase because of their relative delay, $\Delta\tau$. Their frequency difference was used as an estimate of the optical bandwidth of the fiber, which could then be related directly to the bit rate (1/T). For an NRZ baseband signal the 3 dB spectral frequency is 1/2T. The time spread derived this way should be con-sidered an estimate of the width of the fiber's impulse response. Full-width-half-max (FWHM) values were used for both time and frequency domain calculations. For the other two main components, source and receiver, rise time estimates were used. The fiber delay was transformed to a rise time and combined with these to get a system rise time. This simplified approach does not include information on the shape of the pulse, however, which is needed to derive estimates of the bit-error-rate (BER) and the system eye diagram closure. Both of these are important in estimating system perfor-mance and were derived using the ref curves in Figure 7-3 and Equation 7.5, respectively.

A second approach is more mathematically rigorous and includes the effects of pulse shape but was not included in this text. A Gaussian shaped impulse response is often assumed because it closely approximates results in nature, and also allows straightforward BER calculations. In Section

7.2.3 it was pointed out that a Gaussian pulse assumption results in a 3 dB frequency of $0.44/\Delta\tau$ while the intuitive assumption gave a $0.5/\Delta\tau$ estimate. Because the shape of the Gaussian pulse can be described mathematically, any ISI caused by distortions can be modeled directly and the resulting BER calculated. The combined impulse response of the laser + fiber + detector + filter result in a Gaussian shaped pulse.

Assuming a Gaussian output pulse shape and using the $0.5/\Delta\tau$ value for bandwidth results in a 2 dB power penalty. This yields BER curve D in Figure 7.3, which then gives the starting point for the system design in this text. Using equation 7.5, the 2 dB translates to a 37% eye closure. This closure will always be present because the system is dispersion limited. It is not troublesome, however, because the large beginning system margin (31.2 dB) keeps the other noise sources under control. Optical amplifier noise is significant in this system and results in curve E. Curve E is offset from curve D by the 10 dB margin needed to protect against ASE noise. The 21.2 dB power margin that remains (Table 7-6) protects against further deterioration of the BER and eye. Applying equation 7.5, if the eye were to close from 37% to 40% from unanticipated distortion or noise, the margin would lower to 8 dB (21.2–13.2). What that means is, if the signal were to drop to –13 dBm at the detector input (-5 –8), the BER would move up from essentially zero to 10^{-11}. It is not good to operate with a zero margin. A small further decrease in signal (or increase in noise) would move the system quickly into a high error rate condition. The model system used has a good margin which provides an illustration of why digital transmission systems are almost always very good. On the rare occasion when performance deteriorates significantly protection switching is used to change the signal to another path.

These system calculations show problems to be overcome in pushing bit rates to higher levels. The dispersion limitation could be lessened a number of ways: fiber dispersion compensation, faster laser, external modulation, improved receiver response. All of these cost additional money. Also, as WDM system speeds evolve to 10 and 40 Gb/s rates, PMD will become a serious problem requiring compensation. In this system, PMD was found to be only 16 ps which could be ignored.

7.4   Test and Measurement

This section discusses optical and electrical tests and test equipment. They are related to achieving system transmission design objectives in the two areas of power and pulse spreading (dispersion). Some tests do not directly measure power or loss, but measure parameters affecting the power margin. Other tests do not directly measure dispersion but measure parameters that either affect, or are affected by, pulse spreading. Two other categories are also discussed: visual inspection and end-to-end electrical tests. Visual inspection requires only the trained human eye. End-to-end electrical tests assess the combination of power and dispersion effects on a system. The test signals originate at the E/O source and are analyzed after the O/E receiver. Analog signal testing is not discussed. In the same vein, protocol testing of digital signals is also not discussed. This section examines fiber link tests performed at levels data communicators might refer to as the optical and electrical or physical layers.

Fiber system measurements can also be categorized by where the tests are performed. Four environments are identifiable for this purpose: laboratory, manufacturing, field turn-up or commissioning, field maintenance or troubleshooting. This discussion deals with the latter two field-type environments. The measuring equipment used in a laboratory varies markedly from the test equipment used in the field. The two differ in cost and complexity, ease of use and flexibility, ruggedness, size and portability, accuracy and sensitivity. The same optical system parameter may be measured at the two locations, but the approaches will probably be different. Field tests are intended to ensure that a system has been properly designed, installed, and maintained. It is assumed that the channel's basic components (source, optical networking components, fiber, amplifier, and receiver) were provided in accordance with the manufacturer's specifications. Also, it is assumed that the input/output performance of each component is in accordance with these specifications. Most fiber system test equipment is microprocessor controlled and capable of being networked. This affords fast and accurate information processing, data analysis, and data storage.

Still another way to view measuring and testing is through the standards used to ensure industry-wide product uniformity and quality. Standards for measurements and test procedures have been developed in three basic categories: primary, component tests, and system tests. The primary standards give guidance in measuring fundamental physical parameters. Component standards give testing guidelines for fiber optic components, including proper measurement equipment calibration. System standards give measurement methods for point-to point fiber links and fiber networks. System standards would be the area applicable to the tests discussed here.

The tests performed in a field environment are listed on the system block diagram repeated in Figure 7-9. They are identified as being in one of four categories: visual (V), power (P), pulse spreading (S), or end-to-end (E). Tests that can be performed using portable equipment have an asterisk.

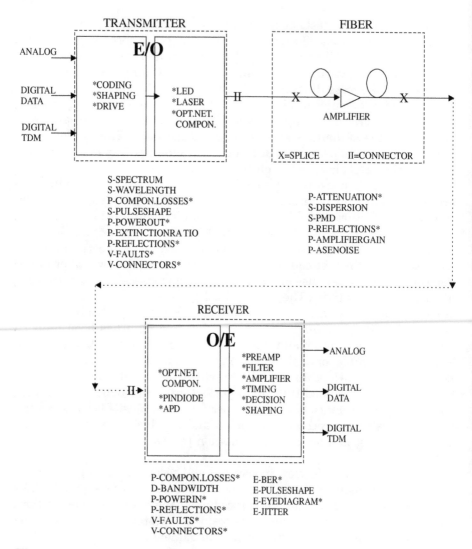

Figure 7-9 *Optical fiber system tests.*

There are optical level and electrical level tests involved. End-to-end tests that have to be made with electrical input/outputs are listed at the receiver. Many of these tests are automated in current test equipment. Some can be controlled from a remote site. This is particularly helpful for DWDM systems used in a metropolitan environment (Chapter 8).

7.4.1 Visual Tests

Visual tests are simple to perform and yet can be extremely valuable to the engineer or technician. A helium-neon (HeNe) laser is used at terminal locations to find sharp bends and defects in fiber jumpers. Helium Neon lasers emit red light that is semicoherent, hence has a narrow linewidth. When coupled into a fiber, some of the light will exit through the cladding/buffer interface at defects or sharp bends and will be visible. Also, the HeNe laser provides a quick check of continuity. Even though attenuation of the red wavelength at 600 nm is high in silica fibers, the light will still be visible at the jumper's output. Safety should always be a concern. The red light emitted from the fiber's end should be illuminated on a surface, not viewed directly.

Another visual test examines connector surfaces with a magnifying device. Hand-held magnifiers are available. The connector's surface is illuminated so that reflected light will reveal imperfections. Connector surfaces should not have cracks, scratches, or identifiable imperfections. Since connectors are sometimes mounted in a field environment, a visual check will be the primary test before more comprehensive optical/electrical tests can be performed. Again, safety requires that people not look into a fiber if the input at the other end is not controlled. The infrared wavelengths used in telecommunications are not noticeable with the naked eye.

7.4.2 Power/Loss Tests

Power output, connector loss, device isolation/insertion loss, and attenuation tests are all made using test sets that contain an optical detector. The optical power meter is probably the most used piece of portable field test equipment. Some power meters will detect multiple wavelengths. Another test set, called the *optical loss test set* (OLTS), combines sources and detectors. Adapters are available so that the many different types of connectors can be accommodated. Calibrated power meters and sources offer the best tools for assessing power levels and losses. Care must be exercised when measuring power in multimode fiber systems because modal fill might affect the measurements. If necessary, mode scramblers can be used to achieve an approximate full-fill condition in a short fiber distance. For single-mode fibers, it may be desirable to use a cladding mode stripper to obtain steady state conditions. This ensures that only the fundamental mode is present in the fiber.

One extremely useful piece of test equipment is the *optical time domain reflectometer* (OTDR), which is basically an optical radar set. An idealized OTDR scan is shown in Figure 7-10. Some of the energy in the main pulse is reflected back at discontinuities. The horizontal time scale gives a reading on the distance to these discontinuities. The velocity of propagation can be set in the OTDR and the return time then translated to distance. The

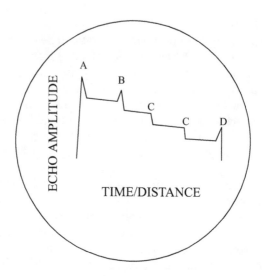

Figure 7-10 *Optical time domain reflectometer (OTDR).*

slope of the trace gives a reading on fiber attenuation. When no disconti-
nuities appear at a particular distance, the energy sent back is caused by
Rayleigh scattering. In high-quality fibers, those with few or no impurities,
Rayleigh scattering accounts for almost all of the attenuation (Chapter 3).
Since the scattering takes place uniformly, the amount scattered back to
the OTDR can be used to measure the attenuation per kilometer. The
sketch shows reflections from the initial fiber connection (A), a connector
(B), and an open end of the fiber (D). The open end could be a broken fiber,
for example. When there is only loss and no Fresnel reflection, the OTDR
does not display returned energy (C). The reflection levels can be measured
at the OTDR, which then allows an assessment of their possible impact on
system performance. OTDRs can be used backwards on only a single span
in an amplified system. Isolators are generally used at the output of optical
amplifiers, allowing only one-way transmission. Fiber amplifiers that do
not have isolators will allow reverse transmission. When fibers in cables
using loose tube construction are measured, the OTDR measured length
may not correlate well with the cable length. Reflections can also be
assessed using the *optical backreflectometer.* It is simpler to use in the field
than the OTDR when measuring only reflections.

Two other methods are used to check fiber attenuation in the field, cutback and insertion loss. The cutback technique is generally preferred
because of the difficulty in accurately measuring the launched power in a
fiber. With the cutback method, the received energy at the far end of the

fiber is measured first. Then the fiber is cut a few meters from the source and the power is measured again. Knowing the length of the fiber, the attenuation in dB/km can then be calculated. The insertion loss method is used for connectorized fibers. The two half connectors at each end are mated and the source and receiver calibrated. The fiber is then connected and another measurement made. The insertion loss divided by the fiber's length gives the attenuation in dB/km. The insertion loss method can also be used on optical components.

Another very useful piece of test equipment is the *optical spectrum analyzer* (OSA). The OSA is the optical equivalent of the conventional spectrum analyzer used with radio systems. Its applications fall into both the power and dispersion areas. The primary uses are to measure source spectra and component bandwidths. The OSA can be used to measure the total power in a signal by summing all the energy in a spectrum. The spectral width and central wavelength of a source are also measured with the OSA. These are important to both power and dispersion assessments. The OSA is also used to analyze the gain shape, four-wave mixing, ASE noise, and S/N of optical amplifiers.

A *multiwavelength meter* (MWM) is used to measure the wavelength of optical sources and their amplitudes. It is simpler to use an MWM for these functions than the OSA. The MWM is also more accurate for measuring wavelengths.

Polarization dependent loss (PDL) is not measured in the field. As discussed in section 7.2.4.5, the PDL of components is in the range of a few tenths of a dB at worst. This loss is included in insertion loss results measured using the OLTS, OSA, MWM, and power meters.

7.4.3 Dispersion/Spreading Tests

Three basic types of dispersion, modal, chromatic, and polarization-mode (PMD) can be measured in the field. For multimode fibers, modal dispersion can be measured by assessing the pulse spread using an oscilloscope with an optical plug-in front end. Another method assesses the bandwidth rather than the pulse spread by using a swept frequency RF signal modulating a light source. As discussed earlier, bandwidth and dispersion are directly related by assuming a first-order rolloff.

Chromatic dispersion is generally not a problem in multimode systems, but has to be measured for long length, high data rate single-mode systems. Early dispersion measuring sets could be used on only a single repeater section because optical sources were not sufficiently narrowband. Also, in amplified systems, the ASE at the receiving terminal would be too great when trying to measure multiple spans. Two different wavelengths are transmitted, each with the same RF sine wave intensity modulation. The difference in phase angle after demodulation gives the group delay and, after calculations, the pulse spread. Newer sets use a narrowband, tunable laser, intensity modulated by the RF signal. Multiple-span, amplified systems can be measured on either a broadband (DWDM) or a single channel basis.

Polarization mode dispersion is important in determining the performance of very high data rate systems (10 and 40 Gb/s). Single-mode fibers suffer from variable birefringence along their transmission path. The two orthogonal polarizations that comprise the fundamental mode arrive at the receiver at different times. The resulting pulse spreading is the polarization-mode dispersion (PMD). Recently manufactured and installed fibers have PMD of < 0.1 ps/$\sqrt{\text{km}}$, but fibers installed before about 1995 are worse. Older fibers need to be measured to determine if higher bit rates (>2.5 Gb/s) can be used. A number of methods are used to measure PMD but some of them cannot be used end-to-end, as desired for a field measurement. The two field methods are the fixed analyzer and the interferometric method. Both arrive at an average *differential group delay* (DGD) result. The fixed analyzer transmits polarized light from a broadband source or a tunable laser. An OSA at the receiver records intensity variations as a function of wavelength. The PMD is derived from the DGD. If an LED is used, about 80 percent of its energy is lost in the step necessary to realize polarized light. A laser's output is already polarized (Chapter 4). In the interferometric approach, the broadband light is separated into two orthogonal polarizations. At the receiver, the delay for one of the polarizations is detected using an interferometer. This method is both reliable and accurate.

7.4.4   End-To-End Tests

Two of the optical tests mentioned earlier can also be performed for the total system, end-to-end: received power level and S/N in each channel of a wavelength multiplexed system. A remote test signal would be launched and power meters or an OSA used to monitor the signal. Any power variation when another channel is added or dropped also needs to be observed. This would be an indication of excessive crosstalk or nonlinear behavior.

Tests of the total system from the transmitter's electrical input to the receiver's electrical output are made using the eye diagram (Figure 7-5) and the bit error rate (BER) curve (Figure 7-3). The eye diagram is displayed on an oscilloscope. BER measurements are made by detecting errors in a transmitted pseudo-random pulse stream, known in advance at the receiver. A BER test set generally has a number of bit rates and patterns available for testing.

8

Future Directions-
Telecommunications

8.1   Introduction

The preceding chapters laid a foundation for understanding individual optical fiber components and how they combine to make point-to-point systems. These discussions basically covered the progress of the technology from the 1970's to the mid-1990's. The systems analyzed in Chapter 7 were relatively uncomplicated, and the 2.5 Gb/s traffic capacity of the digital system was low by current standards. Figure 8-1 shows how progress in two key areas, TDM bit rates and DWDM channel spacing, have led to significant fiber capacity increases in the late 1990's. An individual fiber now has the ability to handle an enormous amount of traffic, provided it has performance characteristics that support the advanced TDM/DWDM electronics/optics equipment. Impressive as these numbers are, for the near future the newest high-capacity systems will probably only be used on a small segment of the international/global network. A problem that affects the usage of new optical technology is a glut of installed fiber, brought about by the excesses of the "dot.com" bubble. Fortunately, the significant optical fiber technology gains generated in the boom years have laid a foundation for future growth. Internet (IP) traffic is still growing at about 50%/yr. It now accounts for a majority of the total traffic on the national network. As a result, the excesses of the boom years should be smoothed out relatively quickly.

For this Chapter, "future" is interpreted to mean the first decade of the new millennium. The period from the late 1990's to 2003 saw a boom and bust cycle for the telecommunications industry. Not by chance, this is the

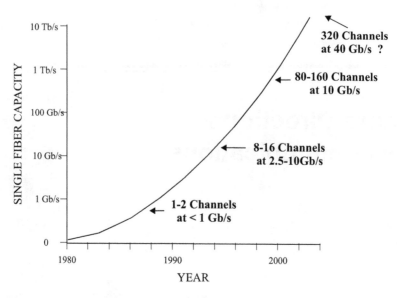

Figure 8-1 *Fiber capacity growth.*

same period that saw the significant gains in individual fiber capacity pictured in Figure 8-1. Economic downturns are always expected and always seem to result in upward corrections in a reasonably short time. The bursting of the "*dot.com*" bubble in the late 90's was an additional negative factor. Both telecommunications service providers and equipment manufacturers suffered greatly with the bust, and many are no longer in business. Experts who analyze this market feel that only about 1–2% of the installed capacity in North America and Europe is *lit*. The term "lit" refers to a fiber that has at least one connected optical source. Submarine cable operators appear to be more efficient, with an estimated 5–20 % of in-place fibers lit, depending on the route. Some routes are being used close to capacity.

The design parameters for a system will determine the ultimate capacity on the fiber being used. The most important of these is the type of (SM) fiber available. For example, is it standard fiber placed in the early 1990's or high quality, dispersion controlled fiber placed 10 years later? Because electronic regeneration is expensive, the distance between O/E/O transformations becomes another important design parameter. This is called the *reach* of an installed system. In the terrestrial network, the designs of *long haul systems* will yield reaches of about 600 km. Submarine cable *ultra-long haul systems* will have reaches in the thousands of kilometers. In an urban environment, *metro systems* will have reaches of about 100 km. Cost is another important design parameter. Metro systems must be less costly per-bit than ultra-long haul systems. Other design considerations include: TDM bit-rate of each channel, WDM channel spacing, optical add/drops, type(s) of amplifiers, amplifier gain and spacing, gain equalization, dispersion compensation, dispersion slope, and PMD.

Regardless of the fiber's quality or characteristics, installed fiber is referred to as *legacy fiber* because it is already there and must be used if possible. Worldwide fiber demand dropped 50% from 2001 to 2002. Dense wavelength division multiplexing (DWDM) equipment, which is used to increase the capacity of installed fiber, experienced a worldwide sales decrease from 2000 to 2002 of about 200%. It appears that the primary objective of the industry in the first decade of the millennium will be to achieve more efficient use of existing fibers. The total *cost-per-bit-per-kilometer (cbk)* will be the important parameter. There will be a combined emphasis on the use of new components, new software/protocols, and systems administration/control to simultaneously meet traffic growth and reduce cbk.

In beginning this discussion of the future it is necessary to recognize the role of networking. Networks are not as simple as they were pre-Internet. Four basic network structures are presented in Section 8.2. Next, two viewpoints of what is basically the same national/international network will be presented in Sections 8.3 and 8.4. Section 8.3 discusses the traditional circuit switched, telephone network usage of optical fibers. This is called the *voice-centric* view of the network. Section 8.4 discusses the packet-switched (Internet) data network usage of the same optical fibers. This is called the *data-centric* view of the network.

This Chapter deals exclusively with high-speed digital transmission using single-mode (SM) fibers. Analog use of optical fibers is relatively limited, and is not driving industry R & D efforts. Multimode (MM) fiber is now used mainly in local area networks (LANs) at short wavelengths. Ongoing R & D is constantly improving SM fibers and optical components. Specifically, significant progress is being realized in the areas of reducing the hydrogen attenuation peak at 1380 nm, optical filtering (DWDM), tunable lasers, wavelength locking, optical amplifiers, chromatic dispersion control, polarization mode dispersion control, optical signal switching, error correction, and wavelength converters. These improvements are discussed in Section 8.5, mainly relative to the performance and reach of single links. Optical networking (Section 8.6) uses some of these to avoid electronic multiplexing/de-multiplexing. With optical networking, wavelengths or "lambdas" are dropped/added/routed through an all-optical network. Connections using only optical paths and/or networking are called *transparent*. A network that uses E/O and O/E conversions is called an *opaque* network.

Networking at the optical layer will be most efficient when accompanied by protocol improvements. As a result, a great deal of R & D is also being concentrated on improving data protocols and packet switching techniques within the framework of a future transparent network. Some background information and possible future directions for protocols are included in Section 8.4. A brief discussion of the importance of international standards is also included in Section 8.4. The standards process has been streamlined compared to earlier years. This has helped speed the introduction of new optical fiber technology.

Finally, coherent optical transmission is discussed in Section 8.7. The concept of optical source coherency was presented in Chapter 4. Currently, coherent optical transmission is not receiving a great deal of R & D attention. Single frequency lasers using wavelength locking and external intensity modulators have an optical spectrum width now governed only by the baseband signal's spectrum. The optical spectrum width is essentially that of a double-sideband AM signal. With DWDM, this is currently a very cost effective approach to the use of available fiber spectrum and is somewhat equivalent to the way commercial RF-AM broadcasts originally used their assigned spectra. In the future the use of coherent systems might allow increased bandwidth efficiency through the application of complex modulation formats (e.g., M-PSK). System costs would be significantly higher, however. As a result, the potential increase in fiber capacity from using coherent systems is not currently needed.

All of the improvements discussed in this Chapter promise three advantages: decreased costs, increased individual fiber capacities, and more widespread networking at the optical level. Since legacy fiber may not optimally match these improvements, designers are facing some interesting challenges. It is ironic that this is the same basic business problem regularly faced by the old Bell System, i.e., the need to shoehorn new, cost-saving technologies and services into an existing, obsolescent network.

8.2   Basic Network Structures

Four common architectures are used to construct digital networks. These are called the bus, ring, star, and mesh (Figure 8-2). The bus structure is generally thought of in connection with Ethernet. Ethernet originally used metallic conductors for 10 and 100 Mb/s transmission. Now, bridges are used to link Ethernets, and optical fibers carry 1 and 10 Gb/s Ethernet. A ring structure example is the Fiber Distributed Data Interface (FDDI) system. FDDI uses counter-propagating fibers at a line rate, with coding, of 125 Mb/s. One fiber provides the primary path and the second a backup. A node or link outage will be detected automatically. Traffic is then

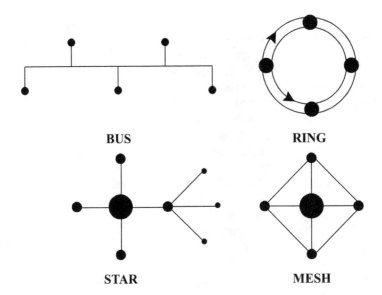

BUS   RING

STAR   MESH

Figure 8-2 *Basic network structures.*

diverted back around the ring using the second fiber. The star is the basic structure used in the telephone network. Telephone switching uses the hierarchy of nodes inherent in a star. Because of the need to make the star more efficient and reliable, alternate path cross-connections are used which then yields a hybrid star/mesh architecture.

Each of these network structures offers unique advantages and disadvantages. The users or clients demand a high level of performance in two areas: S/N, and reliability. Of the four network types, the bus is the most vulnerable and the mesh the least vulnerable to outage. The national telecommunications network is a mesh that uses fiber optic (SONET) rings to provide high performance and reliability.

8.3   The Telephone (Voice-Centric) Network

8.3.1   *Introduction*

The networking of electronic communications can be traced far back in history. As strange as it seems, an extensive digital optical communications network existed in the late 1800s in the American southwest. The network used heliograph or sun-reflecting technology. Using the terminology presented earlier, the heliograph network was a mesh. The Morse code was used for signaling. Network nodes could handle traffic at only about 1–2 b/s. Distances between nodes occasionally spanned hundreds of kilometers. Contrast this with today's optical fiber usage: nodes handle hundreds of Gb/s and can be separated by thousands of kilometers. From Figure 8-1 a single fiber link currently can carry about 1 Terabit/s (T/s) using 80, 10 Gb/s DWDM channels and forward error correction. In a few years this figure will be about 4 Tb/s (80, 40 Gb/s channels).

Initially, transmission networks carried voice or voice band signals in analog format. The first high capacity carrier systems were the analog N and L- carriers used from the 30's through the 70's. They used frequency division multiplexing (FDM). The first digital carrier, T1, began service in the mid- 60's. The T1 line rate was 1.544 Mb/s and was transmitted on twisted wire pair media. A voice signal is converted to 64 kb/s and multiplexed with 23 other voice signals to obtain the T1 line signal. Other signals, like data, were confined to the voice- bandwidth assignment. The T1 baseband signal was labeled DS-1.

The digital networks in the United States and Europe that used early optical fiber systems were designed for telephony-based transmission (voice-centric). Two different digital hierarchies were initially developed for these two regions in support of the *public switched telephone network* (PSTN). T1 became the basis for the *plesiochronous digital hierarchy* (PDH) used in North America until the 90's. The highest line rate in the North American PDH was T5 at 560.160 Mb/s. T5 depended on air-dielectric coaxial cable for long distance transmission, but was transmitted on fiber in the early years of commercial fiber introduction. The European (CCITT) PDH developed line rates similar to the North American rates. The two approaches were sufficiently different, however, that a new hierarchy, SONET/SDH, had to be developed to encourage the growth of international TDM transmission. Also, by developing a new format it was possible to better accommodate data requirements.

To better understand the demand "push" behind new optical fiber technologies, it is first necessary to understand the two distinctly different usages of the same national/international network: a circuit switched telephone network and a packet switched data-com network. Understanding the differences between these two "networks" is important to understanding how optical fiber communication technology is applied and will grow. Even though they use transmission equipment in common, there are significant differences from a traffic and switching viewpoint.

A fundamental difference between telephone traffic and packet-switched traffic lies in their ability to tolerate time delay. The telephone network needs to minimize delay. Speech conversations are troubled when the delay gets beyond about 100 ms. Packet networks are not overly troubled by delay. The use of coding and error correction makes it possible to overcome delay or noise problems. Currently, speech usage of the Internet's packet network is unsatisfactory because of delays, but new protocols will improve performance.

The differences between the two networks lie mainly in the way signals are processed. The telephone network uses electronic switches to connect to the customer and subsequently direct their communications. The data network uses servers and routers to provide these functions. Voice and data traffic have different characteristics. Voice traffic using the telecom network has been analyzed extensively. Voice calls have relatively long holding times; an average of about 3 minutes in the local area and longer for long distance. Also, the amount of voice traffic tends to be equal in both directions. Since talkers are intolerant of delays and breaks in their conversations, it makes practical sense to pair the two directions. The three factors (long holding times, equal traffic in both directions, and no delays) make circuit switched networks more attractive for voice traffic than current packet switched networks.

Data traffic is characterized by short bursts, and is asymmetric. Delays can be tolerated as long as they are not excessive. For example, a residence downloads much more data traffic from its Internet *server* than it originates, and reasonable delays in this transmission are acceptable. On the telephone network data signals are carried either within a 4 kHz wide voice bandwidth or dedicated wider bandwidths. Both become part of a time-division-multiplexed (TDM) stream. As a result, a data signal using the telephone network has to be designed to accommodate existing voice-based transmission standards. Specially designed and installed dedicated (private) links are available at an additional cost.

A fundamental question, integral to the ongoing regulatory process in the US, is whether broadband services in the loop can be met efficiently in a telephone environment. Optical fibers have found only limited application for telephone loops in spite of their large bandwidth capability. *Loops* are the term applied to the short, local connection to the customer. The reasons for this slow penetration are two-fold: relatively high per-customer-unit costs and an apparent lagging demand for high bandwidth capability from residences. *Broadband Passive Optical Networks (B-PONs)* have been developed and deployed to provide high bandwidth capability in the loop area. As the name implies, in a PON active optical components are used sparingly: only at the CO and possibly at a terminal location close to many customers. This results in minimized installation and operating costs but a bandwidth restriction in the metallic drop to the customer. PON systems are currently available for *fiber-to-the-curb* (FTTC) service. Residences then rely on existing twisted pairs to connect to the local node. The next step in extending broadband service in the loop environment is costly, i.e., providing fibers directly to each residence. Installations like this are called *fiber-to-the-home* (FTTH).

8.3.2 *Telephone Hierarchy*

When telecommunications are analyzed from a telephone or voice band viewpoint, the national/international network is best pictured initially as a hierarchy (Figure 8-3). Three basic service areas can be identified in this network: long distance, exchange, and loop. Longer distance links that carry heavy, concentrated traffic are often referred to as *backbone* connections. The final links between the top (regional) offices of the hierarchy are backbones. Lower level offices, in urban environments for example, might also be connected by a backbone because of high traffic volume.

This network structure developed from the beginnings of modern telephony in the early 20th Century and is still valid today — at least from a telephone perspective. In the current jargon, this is a voice-centric view of the national/international network. The network was originally built and controlled by the vertically integrated *Bell System*. Vertical integration meant that the trunks between nodes, as well as the local (loop) connections to the customer, were basically under the control of one entity, the AT&T Company. Long distance trunk service objectives were combined with the exchange and loop service objectives to provide a total, end-to-end *Quality-of-Service (QoS)*. Adherence to high levels of service quality was basically assured through regulatory oversight. In past years new services or equipment added to the telephone network had to conform to the existing telephone network format and objectives. Because of regulatory

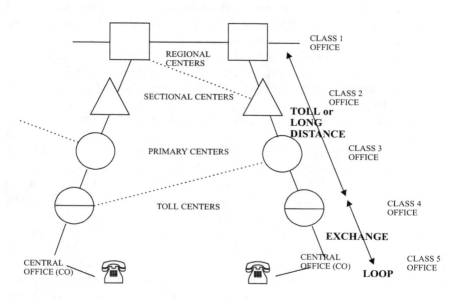

Figure 8-3 *Telephone network.*

restrictions and controls it was somewhat costly and difficult to accommodate radical new services, like Internet packet switching. Ordering the
break-up of this vertically integrated monopoly in 1984 was one approach
regulators felt would help solve this problem. For the Internet/data services network discussed later, QoS is not under the direct control of an
umbrella organization. A QoS objective is negotiated and paid for by the
customer. The Internet protocol in wide use today, does not allow sufficient
QoS service differentiation. Newer protocols are planned that will provide more QoS control and also more efficient usage of DWDM and optical
networks.

The *calling party* initiates the service request and is identified by the
end, Class 5 Central Office (CO). The complete circuit for the call is set up
very quickly through the network switches (nodes). Paths for calls are set
up in advance of call completion using independent signaling channels. The
network is constantly monitored and administered. The voice band connections between nodes are referred to as *trunks*. Trunks are always aggre-
gated through multiplexing to provide more efficient transmission.
Additional nodes and trunks are used to improve network efficiency and
reliability.

As a result, the telephone network can also be described as a hybrid, a
hierarchy/mesh/star network. The links that connect nodes, without going
up and down the hierarchy, are called *tandem trunks*. These provide the
reliability advantages of a mesh network. A hierarchy network alone is vulnerable to blockage and outages when traffic on a link is too high, or a path
broken because of equipment failure. Optical fiber SONET rings are now
used almost exclusively on inter-nodal links because of the need for inexpensive bandwidth and high reliability.

The switching office at the lowest level is called a *central office (CO)*,
and will have many individual customers connected to it. A number of COs
connect to the next level, which is called a toll office. This type of concentration continues up the hierarchy. The connections to the customer from
the CO are referred to as loops. Distances between high-level nodes can be
in the thousands of kilometers, while customer loop connections are generally only a few kilometers. Electronic switches at the network nodes set up
circuit connections, which are simply paths through the network dedicated
for the length of the call. With the TDM terminal equipment currently in
use there is actually not a path in the physical sense. The connection is
sometimes referred to as a *virtual circuit connection*. The signal can be
voice, voice-band data, telemetry, fax, etc. Trunks are installed and administered in pairs carried on the same carrier facilities but serving opposite
directions of transmission. This means that, from the telephone perspective, it is best to have carrier facilities with equal capabilities in both directions to accommodate the two directions of traffic.

8.3.3 *SONET/SDH Multiplexing*

Almost all of the longer distance telephone network links use SONET/SDH-TDM multiplexing and optical fibers. The SONET/SDH standard, discussed in Chapter 7 was adopted in the late 1980's and now has supplanted the PDH hierarchy. The applicable (ITU-T) standard calls for SONET/SDH to be used in rings that provide two routes to any node on the ring. Rings are composed of cascaded, point-to-point, bi-directional links. At each node E/O and O/E conversions are made to allow individual telephone circuits to be dropped or added. The SONET/SDH format was developed to accommodate both voice and data traffic. The original US-T3 (45 Mb/s) rate from PDH is still used extensively, primarily for voice multiplexing. It is converted to the first-level SONET OC-1 (51.84 Mb/s) rate as input to higher speed, optical fiber multiplexed levels.

A SONET/SDH, OC-12, 622 Mb/s multiplexer is pictured as one of the inputs for a 2.5 Gb/s transmitter in the block diagram of Figure 8-4. Because telephone traffic trunks are set up symmetrically, there would be a corresponding demultiplexer paired with the multiplexer to handle the opposite direction of transmission. The system depicted in Figure 8-4 uses WDM to combine multiple services on one fiber. Figure 8-4 also shows the various data inputs that are possible for a high-speed optical carrier. The three public access data inputs, Internet Protocol (IP), frame relay and ATM are shown in one input. This illustrates one of the problems data communications providers want to overcome with new protocols and optical networking. Quality of service levels can be guaranteed with Frame

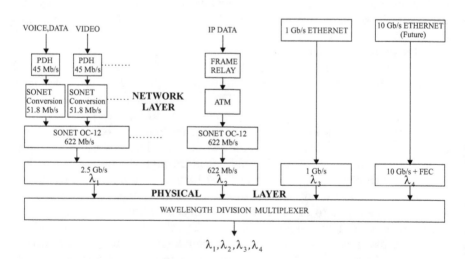

Figure 8-4 *Telephone network inputs.*

relay and ATM. IP is the dominant network protocol and often is fed directly into a SONET/SDH. IP directly into WDM is a future goal of data networkers.

The 2.5 Gb/s final output might drive a laser used for one channel on the WDM system analyzed in Chapter 7. The telephone company OC-12 signal carries 8,064 conversations that have been directed through telephony circuit switches. The *Frame Relay, Asynchronous Transfer Mode* (*ATM*) and *Internet Protocol* (*IP*) blocks identify functions that change serially based data signals to information packets. These might have voice band telephone calls embedded in them. The ATM format is currently more satisfactory for telephone and video signals. The ATM signal might also be carrying IP traffic. The compromises inherent in the SONET/SDH standard made it acceptable initially for both voice and data. Now these are considered to be costly inefficiencies from the data transport viewpoint. Since the volume of data traffic (particularly IP) now exceeds voice-band traffic on the national network, much work is being directed at this issue. Optical fiber networking should help.

Almost all of the transmission/switching on the telephone network is digitally based. The 4 kHz wide analog voice band signals are converted to 64 kb/s digital signals (*A/D conversion*) close to the point of origin. The 64 kb/s signals are time-division-multiplexed (TDM) in the local offices with other signals to form a higher bit rate SONET/SDH stream. The baseband TDM signals are modulated on carrier systems for transport. Signals embedded in the TDM stream are dropped and added at nodes using optical-electrical-optical (OEO) conversion. Data signals are included in the TDM signal. As a result, data signals are forced to accept the framing used by the higher bit rate, TDM line signal. An example here would be communications between two Internet routers at 622 Mb/s (OC-12) using part of a telephone company's SONET connection at 2.5 Gb/s.

For residential customers using wire pairs, telephone companies currently offer *Integrated Services Digital Network* (*ISDN*) and *Digital Subscriber Line* (*DSL*) service. The choice and cost depend on the distance from the CO and the quality of the loop pairs. The ISDN signal speed is 144 kb/s, which is time multiplexed from two 64 kb/s voice channels and a 16 kb/s data signal. The whole 144 kb/s can be used for an Internet connection to provide high speed data service. Digital Subscriber Line comes in different versions. The downstream signal speed can be as low as 1.5 Mb/s or as high as 52 Mb/s. The desired speed determines the maximum length of the customer connection. The upstream bit-rate is usually much lower. An analog voice signal uses the same pair. Since these services originate from a telephone company CO, there are no problems in connecting them to the wider area, high-speed telephone network.

Businesses and enterprises face a somewhat different problem. They are generally willing to invest in upgrading their premise wiring, mainly to improve internal communications. Copper facilities using coax or Category 5 wire pairs (or higher) are available for premise upgrades. These are of significantly higher quality than the standard customer premise wiring, and allow signal frequencies in the hundreds of MHz. Local Area Networks using Ethernet on SM optical fibers can run as fast as 10 Gb/s. Connecting high bit-rate signals to the wider-area, high-speed telephone network becomes a possibly costly issue, however. Competitors to the telephone companies provide connections to the national network from their hubs or head-end terminal locations. These service providers use cable (coax and/or fiber), satellite TV, wireless, and through-the-air optics to reach the customer.

8.4　The Data (Data-Centric) Network

8.4.1　*Introduction*

In the 1980's there was much discussion about a coming *information revolution*. The claim was made that information would, of itself, become a saleable, valuable commodity. Also, that society would become increasingly involved in information generation, transfer and use. Skeptics felt these claims were exaggerated, but in just 20 years they appear to have the ring of truth. The global economy has certainly prospered in the late 20th and early 21st century because of the rapid growth in the data communication infrastructure. The Internet has become the fastest growing traffic category. Methods for achieving electronic communications transitioned rapidly in the 20th century. Now in the US at least, there does indeed appear to be an *infocom* based economy. Manufacturing of goods is being supplanted as the primary force in the US economy by the delivery of services and information through the use of electronic communications. This transition has been made possible by the parallel developments of the microprocessor, integrated circuits, and optical fiber high-speed digital transmission.

8.4.2　*Data Networking*

Data traffic now dictates the growth of the international/national telecommunications network. A data-centric view of the same national network pictured in Figure 8-3 is shown in Figure 8-5. Three basic areas of optical system application are identified: core, metropolitan (metro), and access. The core and metro areas are sketched as rings because of the way SONET/SDH systems are installed. The term "reach" refers to the approximate maximum distance between regenerators. One way to visualize the data network is as an overlay of the telephone network. Much of the transport, terminal, and time-division multiplexing equipment used for tele-

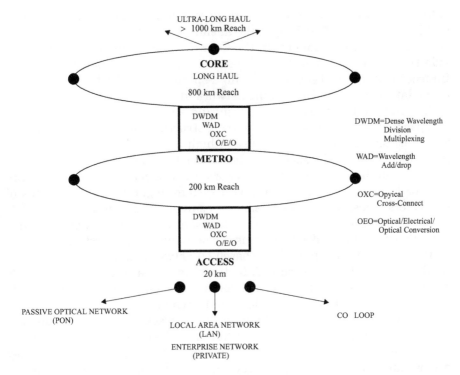

Figure 8-5 *Data network.*

phone service are also used in this data network. The data-centric view of the network is different, however, in the processing of services and the way they are switched. Equipment used in the nodes and to connect to the customer is different. The data-centric, packet network connects to the customer through either a CO loop or access networks.

Two different types of data traffic, unique to modern communications, justify this data-centric view: packet switching, and high-speed (private) data transfers. The former uses the network's SONET/SDH multiplexing for transmission, but routers instead of switches to send packets through the mesh network. The latter uses non-switched data streams, sometimes embedded in higher bit-rate SONET/SDH multiplexed streams. Note that one of the inputs in Figure 8-5 is a Gigabit Ethernet connection. This would possibly be a link for an enterprise or Local Area Network. Gigabit Ethernet and its successor, 10 Gb/s Ethernet, will be important contributors to future optical systems growth in metropolitan areas.

The data transport *Core* network is what the telephone network basically calls the long haul and ultra-long haul toll network. From the data-centric view, the Core network uses a hybrid, mesh/ring/hierarchy architecture. SONET/SDH optical transmission systems in ring configurations are currently dominant in the Core. Telephone companies often provide the OC-48/192 high-speed links that connect Internet routers. The Internet traffic is embedded in the high-speed SONET/SDH signals. Packets of data are moved through the network in a seemingly random fashion, which leads to the need for the mesh/ring structure. Both the telephone and packet networks need the mesh/ring structure for reliability purposes. The hierarchy results from the use of high and low level packet routers, much like there are levels of switches in the telephone network.

Links in the Core part of the network are described as either ultra-long haul or long haul. An ultra-long haul system might extend thousands of kilometers across the Pacific Ocean, for example. The term Wide Area Network (WAN) is used to describe a sub-network within the Core that has long distance links but still retains a community of interest between the nodes. The relationship might be based on network administration/control needs, traffic patterns, or both. A Metropolitan Area Network (MAN) falls between the WAN and access networks. An *access network* corresponds roughly to the telephone local loop and exchange areas. Sometimes the term *edge* is used to describe the interface between the customers and the data-centric network. Local area networks (LANs) are found in the access area. They are sometimes referred to as enterprise networks because they are private or dedicated to a specific customer. Because of competition, the facilities used in these networks are installed and administered by various organizations. SONET/SDH rings are used in the Core and MAN/ WAN networks to provide high bandwidth, traffic flexibility, and assure reliability. With the deployment of 10 Gb/s Ethernet, optical fibers will find increased use in the Access area.

Packet switching is better suited to a data communications network because data traffic is "bursty" in nature. With the circuit switching used in the telephone network, call processing and administration is tightly controlled. In a *packet-switched network* call processing appears to be chaotic, but it is not. To handle the packets, two additional pieces of equipment not needed for the telephone network are added: servers to handle customer connections, and *routers* to perform the nodal switching function. Messages are divided into variable length units or packets and framed for transmission. A header on the packet, placed by the network routers, gives the path to be followed through the network. The maximum length of a packet is fixed by the *protocols*, or rules of data communication. The packets are combined with other packets using statistical concentration, called *statistical multiplexing* and then switched through the network. The path for each packet is currently chosen on an availability basis, which means that some of the message's packets might travel different routes through the network. The receiving terminal reassembles the packets and forwards the total message.

An *Internet Service Provider (ISP)* will accumulate the data inquiries and downloads for the individual customers. The ISP's server corresponds roughly to the telephone network's CO. Internet traffic originates in the form of *datagrams*, which are groupings or packets of data bits sized appropriately by the network protocols. Each packet contains the necessary protocol information. It includes a message sequence number and a destination address. A signal requiring many bits will be broken into multiple packets. The packets can originate from either an ATM or IP switch. The routers can be thought of as the telephone network's nodal switches. The routers store, multiplex, and forward the datagrams- from one router to the next. The packets are sent between routers essentially based on space available. The result is that the packets comprising a single message may use different physical paths through the network. All the packets are reassembled in the proper order at the receiving destination. There is a router hierarchy similar to the telephone switching hierarchy.

Three levels of router service can be used to describe Internet switching. In LANs and enterprise networks, routers can connect thousands of computers. Routers in the access area connect individual residences and small businesses to the Internet. In the MAN, WAN and Core networks, backbone routers link the ISPs and LANs to the long distance network. Some of these backbone routers have ports operating at the high SONET speeds of 10 and 40 Gb/s. The US Internet network uses 12, high level routers. Routers set up virtual connections, not the fixed connections achieved by telephone switches.

8.4.3 *Protocols*

Unlike the telephone network in Figure 8-3, the interfaces between the main sections of the data-centric network are not precisely defined. One reason is that, with the telephone network, specific organizations are responsible for the three main sections: long distance carriers handle interstate toll, and the local Bell companies handle most of the exchange and loop. With the current data network, many different service providers are active at all levels. Another reason interfaces are blurred is because the protocols governing packet communications are layered and initiated/detected at different locations in the network.

The data-centric network relies on this layered, software defined, architecture in addition to its physical architecture. The software layers are not identified by the physical partitions pictured in Figure 8-5. Layers are found within the network's servers and routers or, perhaps, even at a customer's enterprise router. Layers are defined by the inputs and outputs they support, and the function(s) they perform. Communication between equal layer levels in different data systems is governed by the protocol rules.

Protocols are basically the rules used by data terminals to implement the efficient and accurate transfer of data. A data communications network uses a hierarchy of protocols to organize the data flow. Data transmitted in a voice band, embedded in a TDM multiplexed signal will have a separate set of protocols compared to the packet network signals considered here. Packet switching uses multiple protocol layers. The *International standards organization (ISO)* has developed a general description of the way packet network layers should communicate using these layers. This is called the *Open Systems Interconnection (OSI) model*. The OSI recommendation uses 7 layers, as listed in Table 8-1.

The first three bottom layers are of importance to optical fiber system design and deployment. These are the *physical, data link control, and network layers*. The *Network Layer* arranges data packets and sets up the path between terminals. The *Data Link Layer* frames the data and detects and corrects errors. The lowest, *Physical Layer*, specifies hardware parameters. For example, the RS-232 standard is a Physical Layer protocol. Because of the possible advantages with optical networking, a great deal of thought is going into how to best use the WDM layer.

The OSI model gives only general guidelines. Many different protocols have been developed for use at the lower levels of the ISO hierarchy. Optical systems are in the Physical Layer, but optical equipment currently does not use data protocols. Data protocols will play an increasingly important role in the development of optical networks, however. Packet switching at the optical level is one area of technology receiving a great deal of attention. It promises a means of efficiently using optical networking.

Three protocols are currently important in the application of optical fiber systems: frame relay, asynchronous transfer mode (ATM), and Internet Protocol (IP). In a reverse relationship, the proven reliability, quality of

Table 8-1 *Open systems interconnection protocol layers.*

OSI Layer	Name	Characteristics
7	Applications	
6	Presentation	
5	Session	
4	Transport	Transmission Control Protocol (TCP)
3	Network	Internet Protocol (IP)
2	Data Link	Network Protocols (e.g. Ethernet)
1	Physical	SONET/SDH
	WDM ?	

service, and low cost of high-speed optical systems are important in the growth and use of these protocols. These are packet protocols. An earlier protocol, X.25, has been used extensively in the circuit switched, public data networks. It was reliable but slow, and basically designed to accommodate a voice-centric network. The X.25 protocol controlled the data terminal-network interface. It matched the bottom OSI layers. The three newer protocols are not as all encompassing.

Frame relay uses variable-length information packets, and lower *overhead* than its predecessor, X.25. Variable length packets are desirable for data signals because of their "bursty" nature. One drawback with frame relay is that there is less overhead available for error correction. When optical fiber systems are used, this is not a big disadvantage because of their high reliability. In frame relay the maximum length packet, 4096 bytes, can cause storage and synchronization problems, however. A large maximum length packet in a variable length protocol requires more built-in network intelligence, hence higher costs. Further, long packets cause problems for real time services (audio and video) because of the processing delays. These signals are characterized by long holding times and the inability to accept delay. Data signals can tolerate a delay before delivery.

ATM uses small packets of equal size, 53 bytes. This is often referred to as a cell structure. Adopting small packets was a compromise between the needs of the data signals and the real time requirements of digitized voice and data. The small packets result in increased overhead, unfortunately, which makes ATM better suited for high capacity (optical) transmission systems.

The *Internet* is a network of networks using packet switching. It has no central control entity or location. Computers and computer-based networks connect to the Internet through routers. The routers convert any local protocols to the broader network protocol called *Transmission Control Protocol/Internet Protocol (TCP/IP)*. With TCP/IP, data is transmitted in the form of datagrams. A datagram may originally require multiple packets. A packet is delineated by source and destination addresses. The routers use look-up tables to route the datagrams or at least send them in the right direction. The World Wide Web (WWW) is a database system that uses the Internet for transport.

8.5 Point-to-Point/Networking Technology

Point-to-point transmission was the only type of optical fiber system application during the first two decades of optical fiber use (late 1970s to late 1990s). The succession of optical fiber systems, discussed in Chapter 7, was based only on point-to-point applications. Sample designs of links for point-to-point analog and digital systems were studied in Chapter 7. The second example, a digital 1000 km link using optical amplifiers, is representative of high capacity systems used today in the long haul portions of the networks pictured in Figures 8-3, 5. In transmitting from coast to coast, four or five of these links would be connected in tandem. The high-speed digital signal at the nodes would be demodulated, regenerated electronically, and then re-modulated on an optical carrier for the next link. Before the commercial availability of EDFAs in the early 90's, signals had to go through electrical regeneration after a distance of 80–100 km. The term reach is used to describe this distance between regeneration points on a route. If the reach can be extended, the cost (cpk) is lowered significantly. Submarine systems using amplifiers are designed to have reaches of thousands of kilometers. Achieving these distances requires the amplifier sections to be held to about 40 km because of the need to carefully balance all signal distortion contributors.

Dropping and adding lower bit-rate signals on point-to-point links is achieved after O/E conversion and de-multiplexing of the high-speed TDM signal. This is a form of opaque networking achieved at the baseband or electrical level. A communications network occurs when locations with a common interest are interconnected. In the original telephone system, networks could only be realized at the voice-band level. With the widespread use of computers, networking now is associated more with high-speed data connections. Section 8.6 presents some of the future possibilities of networking at the optical level. Optical networking will avoid expensive O/E & E/O conversions, but will require cost reductions in optical components and advances in data communications protocols.

The main objective of fiber and component improvements in the near future will be to obtain the lowest cost-per-bit-per-kilometer (cbk) in both initial system installation and ongoing operations. From experience, the maximum (potential) capacity of systems has doubled about every 18 months and the cpk has been halved about every 9 months. This trend will probably continue for the near future. The explosive growth in data and Internet traffic has essentially dictated that steadily increasing amounts of inexpensive bandwidth be made available. These two demands, traffic increase and cost efficiency, spurred the dramatic improvements in optical fiber technology that took place through the 90's. When these two requirements can be met using portions of the existing optical infrastructure the term applied is *scalability*. An example of scalability would be increasing the bit rate on a transparent connection from 10 to 40 Gb/s.

Point-to-point optical fiber systems and optical networks will make use of technological improvements in the following areas:

- Attenuation optimized SM fiber
- Dispersion optimized SM fiber
- Dispersion compensation
- Broadband optical amplifiers
- DWDM
- Tunable lasers/wavelockers
- Integrated optics/electronics
- Wavelength conversion

All of these help lower the cbk by achieving one or more of the following objectives: adding more channels per fiber, increasing the capacity of a single optical carrier, or increasing the distance between expensive electronic regenerators. Each is discussed in more detail below.

8.5.1 *Attenuation Optimized SM Fiber*

Standard single-mode fiber cannot be used for 10 and 40 Gb/s systems because the minimum attenuation and chromatic dispersion wavelengths do not coincide. Even use beyond OC-12 (622 Mb/s) is confined to relatively short distances. Recall that the wavelengths of lowest attenuation and lowest chromatic dispersion in conventional SM fiber occur at 1550 nm and 1300 nm respectively. The 1550 nm attenuation is at the absolute minimum, which is the Rayleigh scattering limit of 0.16 dB/km. This fiber is referred to now as legacy fiber, and many fiber-km were installed before economical 1550 nm components and systems became available. Transmission at 1550 nm is very attractive because of low loss and wide bandwidth. This allows both increased distances between regenerators and increased fiber capacity. The newer 1550 nm systems employ Erbium doped fiber amplifiers (EDFA) amplifiers and WDM. To gain increased fiber capacity and regeneration distances in the most efficient way, the low dispersion wavelengths must be made to coincide with the 1550 nm low attenuation wavelength. This is achieved with the dispersion-shifted fibers discussed in Chapter 2 and Section 8.5.2.

Recall that standard SM fiber has three water-induced peaks at 1390, 1240, and 950 nm (Chapter 3). The peaks define the three windows of operation used initially for fiber systems that are centered at 850 nm, 1300 nm, and 1550 nm (Figure 3-1). Optical fiber is now available commercially that does away with the water peaks. This is referred to as *dry fiber*. With the additional use of broadband optical amplifiers and WDM filtering, essentially the whole single-mode band between 1300 and 1600 nm can now be exploited. In dry fiber the Silica SM fiber attenuation curve now closely follows the loss limits discussed in Chapter 3 dictated by Rayleigh scattering and infrared absorption.

Table 8-2 shows the bands of wavelengths in SM fiber that have been standardized by the ITU-T. The low window band around 850 nm is used for multimode, LAN WDM systems but is not part of the standard. The C-band was assigned and used first because it coincided with the region of lowest attenuation and the introduction of optical amplifiers in the early 1990s. The "C" designation is used because it is the "Central" or "Conventional" band. Logically, the "L" band label was derived from "Long" wavelengths, and the "S" band from "Short". Use of the "O","E", and "U" bands calls for careful engineering. Wavelengths in the O band are close to cutoff, where multi-moding effects might prove troublesome. The first SM systems used this band. The E band requires the use of dry fiber and has seen only limited use. Also, amplifiers are not commercially available for the O, E, and S wavelengths. The E band will probably find use in the metropolitan and access areas, where amplification is not a primary concern. Some suppliers define the three bands of primary interest, S-C-L, using 5 nm interband spacing. The S band is then 1465–1525 nm, the C band is 1535–1560 nm, and the L band 1570–1620 nm.

8.5.2   *Dispersion Optimized SM Fiber*

Recall from Chapter 3, Figure 3–5, that the standard SM fiber chromatic dispersion parameter D_{chr} has the following key characteristics: a dispersion zero at 1310 nm, dispersion equal to about -18 ns/nm-km in the 1550 nm window. Recall also that the total chromatic dispersion is the sum of two components: material dispersion, and waveguide dispersion. The material dispersion component, D_{mat}, is dictated by the dependency of the core material's index of refraction on wavelength. At any given wavelength this value is determined solely by the chemical composition of the glass. By shaping the core's index of refraction profile, however, the value and slope

Table 8-2   *WDM Bands*

Band	Title	Range (nm)
LOW	First Window	800–1000
O	Original	1260–1360
E	Extended	1360–1460
S	Short	1460–1530
C	Central	1530–1565
L	Long	1565–1625
U	Ultra-long	1625–1675

of the waveguide dispersion component, D_{wag}, can be controlled. By manipulating the waveguide component the wavelength at which the total dispersion becomes zero is shifted higher and the shape and slope of the chromatic dispersion curve is flattened or curved. A number of these core shapes are sketched in Figure 2-12. They achieve different results, but all are referred to as dispersion-shifted fibers. The D_{chr} curve and the zero wavelength will always be shifted to the right.

A word of clarification is in order relative to the way D_{chr} curves are plotted in some other books and references. The relationship of pulse spreading to the dispersion parameter was developed in Equation 3.14 and 3.15 and is repeated here:

$$\Delta\tau = -(\Delta\tau)\,DL_{chr} = -(\Delta\lambda)\,L(D_{mat} + D_{wag}) \qquad 8.1$$

The minus sign stems from the derivation of the arrival time dependence on wavelength in Equation 3.13. Some presentations of the dispersion curve drop the minus sign. This gives the same results as long as the D_{chr} curve is correspondingly inverted. Consider a pulse with all wavelength components above the zero crossing wavelength, for example at 1550 nm. Using the minus sign approach of Equation 8.1 and Figure 3-5, the D_{chr} parameter is negative, thus making $\Delta\tau$ positive. If the minus sign is dropped and the curve inverted, $\Delta\tau$ is positive. When $\Delta\tau$ is positive the shorter wavelengths travel more quickly through the fiber than the longer wavelengths. Either approach preserves this relationship.

Large waveguide dispersion effects are achieved when the fiber's normalized frequency, V, is well below the single-mode cutoff value of 2.405. From Chapter 2, V is directly proportional to the fiber's radius and numerical aperture, and inversely proportional to the wavelength. For a given wavelength, a decreased V is achieved in two ways: by decreasing the effective radius of the core through the use of extra core/cladding layers; and by using a relatively high effective index difference between core and cladding. This has the added advantage of increasing the numerical aperture (Equation 2.10).

The dispersion-shifted fibers used initially in commercial service had a zero crossing wavelength in the middle of the C band. These are now referred to as *zero dispersion-shifted fibers* (ZDS) to distinguish them from dispersion-shifted fibers in which the zero is either above or below the C & L wavelength bands. This latter group is referred to as *non-zero dispersion shifted* fibers (NZDS). ZDS fibers became undesirable when Erbium doped fiber amplifiers (EDFA), WDM, and dispersion-compensating fibers (Section 8.4.3) began to be used. A drawback with NZDS fiber is that it will have slightly higher attenuation.

An NZDS fiber was used for the example system in Chapter 7. The dispersion zero was shifted towards the band of interest from the standard

1310 nm position, but did not fall within the band. From Tables 7-6 & 7, the example system was fiber dispersion limited and this controlled the system's rise time budget. The NZDS fiber had small values of chromatic dispersion and dispersion slope at the four (C-band) wavelengths used for the system. Dispersion-compensating fibers were not needed because the bit rate was 2.5 Gb/s and the dispersion was only 2 ps/(nm-km). If transmission at 10 and 40 Gb/s had been required it would have been necessary to include dispersion compensation because of accumulated delays over the long distance. Recall that the presence of a small amount of dispersion and dispersion slope across the band is desirable. Without some dispersion, four-wave mixing (FWM) could be troublesome because of phase coherence between the WDM channels (Section 7.2.4.5).

The example system in Chapter 7 was designed using the simplifying assumption that all channels were exposed to the same amount of dispersion. This would make the pulse spreading due to chromatic dispersion the same in each WDM channel. In reality, the slope of the dispersion curve across the band of interest can generate important differences between channels. As a result, the dispersion in each WDM channel system has to be considered and the system design should be based upon the worst expected dispersion. Even if an initial design is for only one optical channel, future capacity expansion may require additional channels. In commercial NZDS fiber, the *dispersion slope* is about 0.08 ps/$nm^2 - km$ in the 1550 nm range. In the example system of Chapter 7, the dispersion was assumed to be 2 $ps/nm - km$ for all channels and the channel spacing was 1.6 nm. Taking account of the slope results in a chromatic dispersion difference of about 0.4 $ps/nm - km$ between the end channels of the four. This amount of difference, 20 %, should be taken into account in the total WDM system design. Because the example system was fiber dispersion limited and the NZDS fiber's zero was at a wavelength below the WDM band, the longest wavelength in the WDM group would have the most dispersion. In the example system, designing to the longest wavelength channel means that the other three channels would have slightly better performance. In some NZDS fibers a channel at the top of the C-band can have much greater dispersion than a channel at the low end. As a rule of thumb, uncompensated dispersion accumulation in the channel most affected limits the usable optical bandwidth to about 15 nm at 1550 nm. Reduced-slope fibers are available to keep the dispersion variation across the combined C and L bands to a minimum.

Dispersion-shifted fibers have a slightly smaller (fundamental) mode diameter than standard SM fibers, about 9μm compared to 10μm at 1550 nm. The core index of refraction profiles of dispersion-shifted fibers have rings that increase the concentration of the mode field. Since the core area where the energy concentrates varies with the square of the radius, a small decrease in effective mode diameter significantly increases the core's energy density. This makes dispersion-shifted fibers more vulnerable to non-linear effects. There are core designs that purposefully spread the mode field diameter, but this leads to greater dispersion slope. The effects of four wave mixing are decreased in these fibers, but the dispersion difference across any WDM channels is increased. There never seems to be a "free lunch".

The standard SM "legacy" fiber, with a dispersion zero at 1300 nm, also has a polarization mode dispersion (PMD) problem. Recall from Section 7.2.4.4 that PMD arises when the two orthogonal polarized modes of the fundamental mode experience slightly different delays. Careful control of the fiber's circular symmetry during manufacture minimizes PMD. Dispersion-shifted SM fibers manufactured since the mid 90's can be purchased with low values of PMD ($\approx 0.1\ ps/\sqrt{km}$). Using the example system dispersion of 2 ps/km in uncompensated NZDS fiber, a 40 Gb/s NRZ system would be able to transmit a distance of about 10 km. This would probably lead to the adoption of dispersion compensation techniques (Section 8.5.3). With chromatic dispersion lessened by, say, a factor of 100 through compensation, the PMD would become more troublesome. Over a distance of 100 km the chromatic dispersion would be about 2 ps and the PMD 1 ps, thus becoming a new component of dispersion that might limit system distance.

Unfortunately, the cable environment also affects PMD. Some environmental effects might be static and others dynamic and unpredictable. As a result, if bit rates > 2.5 GH/z are planned, PMD measurements must be made on the installed fiber. Commercial PMD compensators are not currently available. If fiber installed before the mid-90's shows high PMD, as it probably will, the only alternative is to shorten the length between regenerators. This is the main reason that 40 Gb/s systems will probably find initial use in the Metropolitan and Access areas of the data-centric network. The distances between nodes are shorter than those in the Core area.

8.5.3 *Chromatic Dispersion/Slope Compensation*

Dispersion compensation is achieved by correcting for the length-dependent chromatic dispersion of a fiber. Either distributed or lumped compensators are used. When 10 Gb/s systems were first introduced, the effects of dispersion were found to limit performance and a "quick fix" was required. Dispersion compensation module "add-ons" using fibers were adopted, and are still used as the primary method of dispersion compensation.

A large amount of SM dispersion-shifted fiber, with the zero in the middle of the third window, was installed in the 1990's. The initial thinking was to accommodate point-to-point 1550 nm systems or possibly 1300/1550 nm, dual channel WDM combinations. Like the standard SM fiber, this fiber is now considered legacy fiber because it is not optimum for DWDM systems.

With DWDM it is necessary first to use dispersion-shifted fiber but, second, that the fiber's zero not be in the WDM band(s) of interest. As both the technological capabilities and the demand for bandwidth have increased, the ability to combine C and L band WDM systems on dispersion-shifted fibers has become important. In addition, increasing bit rates from 2.5 Gb/s to 10 or, ultimately, 40 Gb/s has also become very desirable. The use of very high bit-rates on closely packed wavelengths requires that both chromatic dispersion and dispersion slope be minimized through corrective action.

Chromatic dispersion is currently compensated passively, not actively, using two basic types of compensation: lengths of specially designed fibers called *dispersion compensating fiber (DCF)*, and fiber Bragg gratings. In general, lengths of fibers can compensate over a broader range of wavelengths than Bragg gratings. As a result, in a DWDM grouping that uses fiber compensation, only one wavelength can be accurately optimized. The other channels will have significant delay improvement, but residual delay differentials will exist between wavelengths. Also, it is impossible to exactly compensate for dispersion slope, which further adds to delay differences between channels. Dynamic compensation will eventually be available that will allow automatic control of dispersion compensation. Compensation can then be specifically adjusted for each channel in a WDM group. These types of compensators will probably be relatively expensive. They will find primary application in optical networks where wavelengths are routed individually.

Many combinations of long length, primary fiber and shorter compensating fiber are possible. Two main approaches are illustrated in Figure 8-6. Figure 8-6A sketches an approach that concatenates approximately equal lengths of fiber with opposite dispersion signs and slope. This could be accomplished as shown by cascading sections of NZDS fiber with zero wavelengths alternately above and below the band of interest. Since both types of fiber work together, a specific DCF fiber is not identified. If slope dispersion correction is included, the fibers would have carefully designed cores. A combination like this must be planned in advance of installation. Submarine cable installations use this approach, which results in custom tailored chromatic dispersion over very long distances. Submarine cable repeaters now use optical amplifiers only in mid-span. Dispersion compensation allows thousands of kilometers of fiber to be used between electronic signal regenerators. Terrestrial systems do not have a similar need to stretch the reach, thus their designs are not as demanding.

Figure 8-6B shows a short dispersion compensating fiber (DCF) loop or module used to compensate a previously installed longer fiber. This might be an application that upgrades legacy fiber for a higher speed in a metro area. The primary objective in this example is to improve dispersion performance across the C-band. DCFs generally offer 80 to 100 ps/(nm-km) of

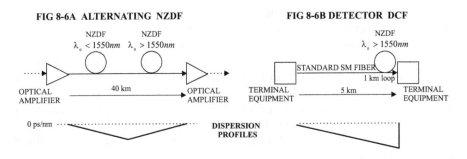

Figure 8-6 *Dispersion compensating fiber.*

compensation in the C-band. As an example, the DCF could be a NZDS fiber with a zero beyond 1600 nm and a large positive dispersion of 90 ps/(nm-km) at 1550 nm. The main fiber could be a standard SM fiber with −18 ps/(nm-km) of dispersion at 1550 nm and a zero at 1310 nm. In this example 1 km of coiled DCF fiber would be used to compensate for 5 km of the previously installed standard SM fiber. In Figure 8-6B the DCF is placed at the receiver where the energy density will be lowest. NZDS fibers have smaller effective fundamental mode area, hence higher core energy densities and increased susceptibility to four-wave-mixing. Their attenuation is higher than the standard fiber, which must also be taken into account.

Optimally, dispersion compensation should restore pulses carried on each wavelength to their original (time domain) shapes at the detector. Paradoxically, an important objective in planning dispersion compensation is to keep the dispersion value nonzero throughout the fiber's length. Ideally, only at the detector should it be brought to zero. Recall from Chapter 7 that the effects of four-wave-mixing (FWM) are made worse when WDM carriers are in phase. The presence of dispersion and dispersion slope at all points on the fiber's length, except at the detector, disrupts this phase correlation and minimizes FWM interference.

The dispersion profiles included in Figure 8-6 for a single channel are useful in assessing the effectiveness of a dispersion compensation approach. A *dispersion profile* gives a plot of expected dispersion as a function of distance along the fiber. A companion *power profile*, that gives expected power levels on amplified fiber sections as a function of fiber length, can also be developed. The pair provides very useful engineering design information.

Continuous compensation over a broad wavelength range can also be achieved by inserting *high-order-mode* (*HOM*) fibers. An HOM fiber propagates the fundamental mode plus selected higher order modes. From Chapter 2, higher order modes travel at slower group velocities. By designing the coupling to be wavelength sensitive, the total time delay of the pulse can be compressed. This basic approach using differential delays is also achieved in fiber Bragg grating compensators.

Bragg gratings find many applications in advanced fiber optic components. A Bragg grating was discussed in Section 4.4.4 in obtaining single-longitudinal mode output from DFB and DFR lasers, for example. A Bragg grating in a fiber essentially provides an optically tuned circuit filter. It is much like a filter made of thin films, except that the layers are sections of fiber with alternating sections of high and low refractive index. The layers are achieved using external radiation to change portions of the fiber's refractive index.

A fiber Bragg grating compensator is sketched schematically in Figure 8-7. Assume the four input wavelengths are above the dispersion zero wavelength. That means the longest wavelength is the slowest through the fiber and has arrived later than the shorter wavelengths. The chirped fiber Bragg grating will have to provide compensation in the opposite direction, i.e., slowing the shortest relative to the longest.

The four input wavelengths enter the circulator at port 1 and exit into the Bragg grating from port 2. The circulator separates the grating's input from the dispersion compensated output. The output is a combination of reflected waves that have traveled different distances into the grating. The distance between higher index layers, plus their thickness, determines the reflected wavelength at a particular point in the grating. Wavelengths that do not match the grating period(s) are transmitted without reflection. This grating is chirped, which means that different wavelengths are reflected at different positions in the grating. In this figure the longer wavelengths are reflected before the shorter wavelengths. This compensates for a $\Delta\tau$ caused by the shorter wavelengths traveling faster through the fiber. As a result, the pulse at the circulator's output is compressed to a width closer to the original.

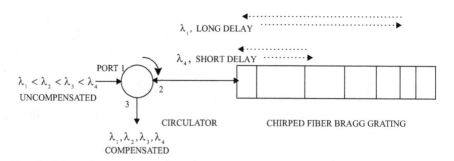

Figure 8-7 *Fiber bragg grating compensation.*

In the 2.5 Gb/s example system in Chapter 7, the chromatic dispersion was —2 ps/(nm-km). This made $\Delta\tau = 0.2ns$. Because it is positive, the shorter wavelengths are traveling faster than the longer wavelengths. To compensate for the $\Delta\tau$ time difference in length between the long and short wavelength reflectors, the chirped grating in Silica would have to be about 20 mm long. It is difficult to manufacture long fiber Bragg gratings, so current commercial fiber gratings are limited in total length. At 1550 nm, one chirped Bragg grating will be able to compensate for about 5 nm of bandwidth. Gratings have to be used in tandem if broader widths are required. This can be a problem since insertion losses, including the circulator, are a few dB for each compensator.

8.5.4 *Broadband Optical Amplifiers*

If there is a single optical technology responsible for the rapid growth in optical fiber use it is probably the optical amplifier. Optical amplifiers made it possible to push out the distances between electrical regenerators to thousands of kilometers. Recall that the electrical regeneration process reconstructs each digital pulse. This requires costly O/E and E/O conversions. Also, the necessary electronics had to be dedicated to each optical carrier. Another advantage with optical amplifiers is that they are transparent to the line speed modulated on the optical carrier. This saves money and improves reliability, especially for submarine cable systems. As an aside, submarine cable jargon refers to optical amplifiers as repeaters, even though the term *repeater* normally means a back-to-back receiver and transmitter pair. Also, the broad bandwidths of optical amplifiers have made DWDM and optical networking economically attractive.

Four basic types of optical amplifiers are now commercially available: Erbium-doped fiber amplifiers (EDFAs), *Erbium-doped waveguide amplifiers* (EDWAs) *Raman amplifiers,* and *semiconductor optical amplifiers* (*SOAs*). Table 8-3 lists some of their characteristics and applications:

The first optical amplifier in commercial use was the EDFA. It was introduced in the early 1990s for use on a single 1550 nm channel. EDFAs still account for the bulk of the optical amplifier market. EDFAs became popular because of their ability to amplify many WDM channels simultaneously. In the C-band, for example, an EDFA amplifier might be transmitting 80 channels with 50 GHz spacing (0.4 nm at 1550 nm). EDFAs are used for amplification in both the C and L bands, a total spectral bandwidth of 95 nm between 1530 and 1625 nm. The two bands require separate amplifiers, however. The gain of Erbium doped fiber per unit fiber length is less at longer wavelengths. If a fiber is transmitting both bands simultaneously the bands must first be split, then amplified and recombined.

From Chapter 4, EDFAs use lengths of doped fiber in which energy is transferred from excited Erbium atoms to the signal(s) being transmitted

Table 8-3 *Optical Amplifiers*

Name	Bandwidth	Characteristics
Erbium-doped fiber amplifier (EDFA)	C & L bands 1530–1625nm	*Broadband *Need long fibers *ASE noise w/o signal *Low noise figure (NF=3dB) *Polarization independent *Efficient *Inexpensive
Erbium-doped waveguide amplifier (EDWA)	C & L bands 1530–1625nm	*Single channel *Can be integrated *Requires I/O coupling
Praseodymium-doped fiber amplifier (PDFA)	O band 1280–1340nm	*Requires non-silica fiber *high power non-standard pump
Semi-conductor optical amplifier (SOA)	0.8 Window 1300–1600nm	*Can be integrated *Higher noise (NF=6–9dB) *Polarization dependent *Efficient *Inexpensive
Raman amplifier	1300–1600nm	*Distributed (long fiber) or lumped *Gain-balancing w/EDFA *Low noise *Polarization dependent

through the fiber. The erbium atoms are pumped with energy at wavelengths of 1480 nm and/or 980 nm. Pumping can be achieved in either or both directions through the doped fiber. Forward pumping with 980 nm provides a high number of excited atoms and a low noise figure. Backward pumping with 1480 nm achieves a higher output power. The laser pumps are located in the same module as the doped fiber. The pumping signals are added and extracted from the main fiber using wavelength sensitive couplers. EDFAs will saturate at relatively low output power levels of about +20 dBm. As a result, even though they can provide high gain, the output power-per-channel in DWDM configurations will be limited.

Instead of coils of Erbium doped fiber, an EDWA uses Erbium-doped waveguides embedded in substrates. This allows the development of an optical integrated circuit amplifier that is small and uses low cost components. The EDWA will have lower gain than the EDFA because the path length available for energy transfer is shorter. This is not a disadvantage in MAN and Access network applications where distances between nodes are limited, and smaller amplifier gains are satisfactory.

Unfortunately, Erbium based amplifiers do not provide equal gain across all wavelengths. Their spectral gain shape depends on a number of factors: input signal power, pump power, length and doping concentration in the fiber. The gain in an EDFA, C-band amplifier peaks at about 1535 nm, with a variation across the C-band range of about 2 dB. If 5 amplifiers were used in tandem, the result could be an unacceptable 10 dB difference between channels in a full DWDM configuration. Gain compensation is employed through the use of variable attenuators and optical filters. These passive devices still can introduce a systematic ripple across the pass band. An analysis shows that a 2 dB ripple will degrade some of the optical S/N by 3.5 dB.

Transcontinental submarine fiber systems are particularly sensitive to the need for gain shape control because regenerators are far apart. As a result, in submarine systems individual amplifier gains are low. The spacing between amplifiers is on the order of 40 km compared to hundreds of kilometers on land. The close spacing also helps control noise by keeping gain low and input power high.

Another way EDFA gain shape is compensated is through the use of *hybrid amplifiers*. A hybrid amplifier combines Raman amplification with EDFA amplification in a synergistic relationship. The transmission fiber becomes the gain medium. The Raman amplifier is located at the same point in the system as a (low-level) EDFA. Raman gain has a spectral response curve that compensates for the drop off in gain with wavelength that occurs in EDFAs. Hybrid amplifiers have a number of advantages: no insertion loss from gain-equalizing filters or attenuators, noise from distributed Raman amplifiers is relatively low, no specially doped (high-loss) fibers needed. Raman amplification can be used across the total band from 1200 to 1600 nm.

Raman amplification is obtained by using the non-linear, stimulated Raman scattering (SRS) response in a fiber. When a high energy, counter-propagating pump signal is launched into an ordinary Silica fiber at one wavelength, SRS will transfer energy to a higher wavelength signal. In SRS, one wavelength amplifies light at a second wavelength through molecular vibration. This is a frequency dependent phenomenon that depends on the fiber's chemical composition. In Germanium-doped Silica fibers the frequency difference between the two wavelengths is 13.5 THz (about 100 nm wavelength difference at 1550 nm). Germanium has a higher Raman response than pure Silica. Raman amplification has a shape that peaks about 100nm above the pump when the wavelength to be amplified is 1550 nm. The region of useable gain from each pump is about 30 nm wide and falls off linearly from the peak back to the pump wavelength. Gains from the Raman effect alone are kept to about 20 dB in order to minimize added noise. A C-band hybrid amplifier uses a pump at about 1460 nm.

Raman amplifiers have drawbacks. The pump energies have to be high, about 500 mW. This presents safety concerns and the possibility of added nonlinear response in the transmission fiber. The laser pump signal has to be de-polarized because the Raman effect is polarization sensitive. Recall that the incoming signal can have random polarization. A discrete or module Raman amplifier requires a large package or footprint because of the need for a long length of coiled fiber. An advantage here, however, is that special high Germanium fibers can be used even though their attenuation is high. Another potential problem arises if S-band channels are used with C and L band channels on the same fiber. The long wavelength pump signals will interfer with the S-band channels.

Semiconductor optical amplifiers (SOA) are the third main type of amplifier currently in use. An SOA is basically a laser with suppressed internal reflection capability. If the SOA is not in an integrated unit, the natural facet reflectivity has to be diminished with anti-reflective coating. Diminishing the reflections controls added noise and possible instabilities. Noise figures for SOAs are typically 6 to 9 dB. Output power at saturation is also low, about 13 dBm. EDFAs are currently more widely used than SOAs. SOAs have inherent advantages that will become more important in time: they can be used in a broad range of wavelengths (1280 to 1650nm), they have low power consumption, low cost, small footprint, and high speed of response. SOAs will become very important with the anticipated increased use of optical networking. One main advantage of SOAs is that they can be easily integrated into small physical packages. This helps overcome the insertion losses from optical networking devices like star couplers, optical switches or cross-connects, and wavelength converters. They are also useful as modulators because they respond rapidly. They are not used with analog signals because their gain is sensitive to signal intensity, which leads to distortion. Another drawback is sensitivity to the incoming signal's state of polarization. Also, when used as an in-line transmission amplifier, they can present significant coupling losses. SOAs as transmission amplifiers will probably find extensive use in the Metropolitan and Access areas of the data-centric network.

Other amplifier types are being researched and may emerge as contenders to EDFAs, EDWAs, and SOAs. Praeseodymium-doped fiber amplifiers are being developed for use in the 1310 nm area. Thulium-doped fiber amplifiers might also become available for use in the S-band.

8.5.5 *Coarse/Dense Wavelength Division Multiplexing (CWDM/DWDM)*

The commercial availability of C-band Erbium-doped fiber amplifiers paved the way for the use of densely packed wavelength carriers. WDM was used in the 1980s, but only with widely separated wavelengths. With traffic demands growing rapidly in the mid-1990's it became obvious that a progression to very dense packing of individual wavelengths would be required. Figure 8-8 outlines the standardized long wavelength assignments for SM fibers from the ITU-T, G.692.2 (CWDM) and G692 (DWDM) recommendations. The DWDM channel spacing and assignments are based on frequency, not wavelength. The standard was generated by representatives from the telephone industry and, as a result, is more correctly described as frequency division multiplexing (FDM). A more practical reason for using a frequency based reference is that wavelength is affected by the index of refraction of a material. Also, because of the fundamental physical phenomena that governs the radiation, optical sources emit based on frequency not wavelength.

The DWDM assignments are referenced to 193.1 THz (1552.542 nm). The first of 80, 50 GHz channel assignments in the C band has a center frequency of 196.1 THz (1528.77 nm). The next channel has a center frequency 50 GHz lower (0.39 nm higher). For spacing of 100, 200, or 400 GHz, the channel center frequencies are decremented accordingly, starting from the initial starting frequency. Figure 8-8 shows 100 GHz spacing that extends into the L band. To lessen component costs, some DWDM systems are

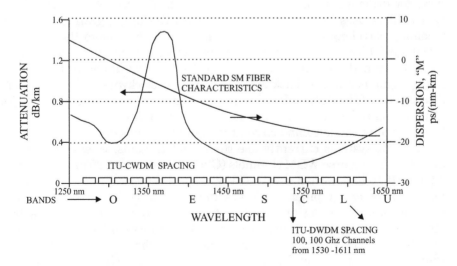

Figure 8-8 *Wavelength division multiplexing.*

designed to interleave channels. An interleaving filter will separate closely spaced wavelengths into two sets of channels that have twice the original spacing.

Systems with channel spacing as narrow as 25 GHz are under development. Achieving channel packing closer than 100 GHz requires that the laser source have exceptional central wavelength and linewidth control. This is achieved with temperature control and external modulators, both of which add costs to the transmitter package. Transmitting bit rates at 40 Gb/s will require $\Delta\lambda$ to be less than about 1 GHz (0.002 nm at 1550 nm) to control chromatic dispersion. In addition, tunable lasers are needed to make DWDM cost effective. It is expensive to stock many fixed wavelength lasers.

The tighter the channels are packed, the greater the cost of separating, combining and isolating the signals. Optical WDM devices were discussed in Chapter 6. The filters required to achieve DWDM must have steep skirts to reject interference, and relatively flat pass-bands to avoid distortion. A typical requirement for the rejection of adjacent channel energy is 30 dB at the adjacent pass-band. A typical pass-band requirement for a 10 Gb/s signal is that the spectral response should be flat to within a few tenths of a dB. The filters also have to be protected from environmentally induced drifting. Two basic types of filters are currently favored: transmission filters that use deposited thin films, and reflection filters that use fiber Bragg gratings. Thin film filters are less expensive. Bragg grating filters provide sharper spectral response, but are more sensitive to the environment. Micro-optic filters are more expensive.

Figure 8-9 shows a block diagram of a WDM system that uses a Bragg filter for dropping/adding a single wavelength. The *wavelength add/drop (WAD)* filter uses two circulators along with the Bragg grating. The four incoming wavelengths to Port 1 circulate clockwise through the left-hand circulator C1.They exit C1 at port 2 and enter the Bragg filter. Wavelengths λ_1, λ_2, λ_3, transit the filter, enter C2 at port 2, and exit at port 3. The incoming λ_4 signal is reflected back to C1. It circulates clockwise and exits at port 3 where it will be demodulated. A new λ_4 signal enters C2 at port 1, exits at port 2, is reflected back and exits C2 at port 3. If a wavelength were to be added only, the left hand circulator would be dropped.

DWDM systems currently use the C and L bands, where EDFAs are available. The coarse WDM channels (CWDM) are spaced 20 nm apart from 1270 to 1610 in accordance with ITU standard G.694.2. This wider spacing leads to smaller equipment packages and lower costs. Laser sources do not need expensive cooling and central frequency control. Filters have less stringent requirements. Inexpensive thin film filters can be used since the pass-band can be as wide as 13 nm of the 20 nm channel assignment. Note that the use of the O and E band channel assignments will probably require the use of *zero-water-peak fiber* (ZWPF). Currently, CWDM systems are

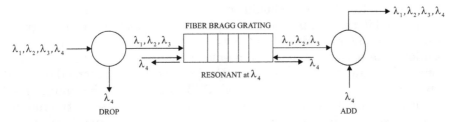

Figure 8-9 *Wavelength drop/add filter.*

designed without amplifiers. This limits their reach to about 80 km. They find primary use in the metro and access areas.

Ethernet at 10 Gb/s will probably be a primary user of CWDM. The Ethernet system has become the standard for local area networks (LANs). Ethernet at 1 Gb/s is transmitted up to 50 km on SM fiber. New 10 Gb/s Ethernet standards have recently been adopted which allow two approaches: the use of four, 3.125 Gb/s (w/coding) signals at 1310 nm in a CWDM configuration, or a single 12.5 Gb/s signal. The fiber can be 62.5/125 MM for short distances or SM fiber for longer distances up to 10 km. Fibre Channel at 10 Gb/s uses 8 CWDM channels spaced 12 nm apart in the 850 nm window on MM fiber. Costs are low because the systems use GaAs VCSELs.

Initially, DWDM was used mainly in the long haul, heavy traffic, core area of the network. The extra costs associated with this method of increasing capacity were more readily absorbed on backbone routes. Now, DWDM systems are being made economically attractive for the metro area. Optical component costs are dropping, and there is a growing recognition of the value of optical networking. Optical networking requires a multiplicity of optical carriers.

8.5.6 *Tunable Lasers*

Wavelength tunable lasers have been used in optical test equipment for many years. For that application, package size and cost are not major concerns. For telecommunications applications, on the other hand, size and cost are extremely important. There are a number of ways diode lasers can be made tunable and still meet these two basic requirements. In addition, tunable diode lasers offer other advantages, all of them associated with implementing inexpensive DWDM systems and optical networking:

- Rapid replacement of failed units
- Add/drop wavelength control
- Rapid reconfiguration/provisioning
- Inventory control

Ideally, a tunable laser should be stable, reliable, scalable (able to handle higher bit rates), and easy to deploy. The package size, or footprint, should be standard and the center wavelength adjustable from a remote location. The package may be stand-alone or integrated with other functions such as an external modulator. For long-haul applications the output power should be at least 10 mW for all DWDM channels. The tuning speed can also be an important parameter, depending on the use. If the tunable laser is a spare to be inserted by a repair technician, the tuning speed can be slow. If the tunable laser has to respond quickly, say by remote control to a system outage, the speed should be fast to avoid loss of too much data.

The most obvious approach to achieving an optical source with selectable wavelengths uses multiple lasers on the same chip. This *arrayed laser* is a grouping of fixed-wavelength, DFB lasers with their outputs combined in a power combiner/splitter. The desired laser in the group is independently modulated on/off. This form of wavelength tuning is coarse and relatively slow.

For a diode laser to be tunable the active medium must have high gain over a wide range of wavelengths. A tunable diode laser that covers the 80 channels in the C-band needs a tuning range of 35 nm. For the 100 channels in the L-band this would be 60 nm. Tunable lasers able to accommodate this wide a band are referred to as *wide tuned lasers (WTL)*. Most tunable diode lasers currently available have narrow tuning ranges, covering from 4 to 20 wavelengths. These are referred to as *narrow tuned lasers (NTL)*. As might be expected, there can be significant cost differences between the two. Erbium-doped fiber lasers can also deliver a wide gain-bandwidth product, but development to date has concentrated on diode lasers.

A tunable laser is a variable oscillator. All oscillators must have three internal capabilities: gain; feedback; and a resonance or frequency sensitive element. In a standard Fabry-Perot diode laser the active region provides the gain and the cleaved facets provide the feedback. The frequency/wavelength selection comes from the length of the diode's active region. To make a tunable laser the cavity length must be variable. How the cavity length and, hence, wavelength variation is obtained distinguishes the two basic types of tunable diode lasers. One approach uses same-chip or monolithic construction. The other uses an external cavity that expands or contracts, thus changing the size of the resonance element.

Currently available monolithic tunable diode lasers use three basic tuning approaches: index tuning, integrated cavity tuning, or arrayed, multiple-wavelength lasers on a single chip. Index tuned diodes use distributed feedback (DFB) or distributed Bragg (DBR) lasers. These are NTLs because their range is relatively limited. Recall that with the DFB laser, the grating that determines the center wavelength is deposited in parallel with the active region. With the DBR laser the Bragg grating is external to the active region. A laser that uses gratings will have a center wavelength determined by the spacing and refractive index of the grating. The spacing and index must be stable if the laser is intended to have a fixed output wavelength. Changing the current in the DFB or the temperature in the DBR varies the index of refraction of the grating and the resulting wavelength of oscillation. The DBR laser can be tuned over a range of 10 nm, greater by a factor of three relative to the DFB laser.

An externally tuned diode has three basic optical components: the laser diode chip; a lens; and a grating/filter. A resonant cavity is formed between the reflecting surface of the grating and the far end of the diode chip. The lens is positioned in the cavity between the filter and the laser diode. The angle of the filter relative to the chip changes the effective cavity length and hence the oscillating wavelength. A motor controls the angle of the filter. This type of tunable laser diode has a broad range but relatively slow response. The output power can be high.

Another type of external cavity tunable laser diode uses VCSELs (vertical-cavity surface-emitting lasers). These were discussed in Chapter 4. Recall that VCSELs have a significant cost advantage because they are manufactured like LEDs. To make a VCSEL tunable, a MEMs mirror is placed at the top of the vertically aligned active region to form an adjustable cavity. Tunable VCSEL sources are relatively slow. VCSELs have been available in the 850 nm window for some time. Primarily because of their cost advantage over edge-emitters, there is a great interest in extending the VCSEL concept to longer wavelengths.

Still another variant of the external cavity approach uses a phased-array-waveguide-grating (PAWG), discussed in Chapter 7. Recall that a PAWG has M input ports and N output ports. N combined wavelengths on one of the input ports are separated and sent to each of the output ports. The PAWG is essentially a grouping of Mach-Zehnder interferometers. Adjacent waveguides in the PAWG provide a wavelength sensitive path to one of the "N" output ports. To make tunable diode lasers with the PAWG, each of the input ports has an optical amplifier to provide gain. Mirrors on the amplifier input and the PAWG's output provide the resonant cavity. Like the laser array, the outputs are at discrete frequencies. PAWG based tunable lasers are available commercially in small packages because of the use of integrated optics.

8.5.7 *Integrated Optics/Electronics*

Optical amplifiers, and the rapid expansion in Internet traffic, spurred the development of DWDM. In similar fashion, DWDM and the Internet are spurring the development of optical and electro-optic integrated circuits. As mentioned, a great deal of fiber capacity was installed in the late 1990's and is still unused years later. The emphasis in the first decade of the new century is on using these fibers as efficiently as possible. Integrating multiple electronic/optical functions in a single package or on a single chip significantly decreases the cbk and helps meet that objective. Low costs are obtained primarily in the manufacturing phase, and result from the use of existing Silicon or GaAs facilities. Low cbk is particularly attractive to service providers in the shorter distance, metro and access service areas. In the future, systems designed to satisfy that market are going to use a great deal of integration.

The number of functions realizable in an integrated optical circuit currently is less than five. Theoretically this number can be much higher. The multiple functions are interconnected using optical planar waveguides embedded in substrates. These are called *planar lightwave circuits* (*PLC*) and have been in use since the 1970s. Early PLCs used Lithium-Niobate (LiNbO$_3$). Lithium-niobate is still used for one class of optical IC. Lithium-niobate exhibits a linear electro-optic sensitivity called the *Pockels effect*. Waveguides are formed on the substrate by diffusing hydrogen or titanium into narrow stripes to raise the index of refraction. An applied electric field changes the index of refraction within the waveguide and thus the velocity of propagation. This change in velocity can be used to make a number of devices: modulators; switches; and polarizers. The electro-optic effect in Lithium-niobate responds rapidly, so modulators at bit rates up to 40 Gb/s and beyond have been realized. A lithium-niobate modulator using a Mach-Zender interferometer is sketched in Figure 8-10. The electric field varies in accordance with the baseband digital signal. The phase of the optical carrier in the upper arm is changed relative to the lower arm because of the difference in propagation velocities. At the output coupler the two signals will either add or subtract, depending on their relative phase shifts. An external modulator is used for high bit rate signals because directly modulated lasers have extinction ratio related noise and distortions caused by chirp.

Currently, the most popular approach for realizing planar lightwave circuits uses silica deposited on silicon (SiO$_2$ on Si). Most photonic ICs are manufactured using the proven techniques of semiconductor IC fabrication: computer design; photolithography; wafer-scale production, testing, and processing. This allows the manufacture of low cost, high quality devices. Even customization similar to electronic chip ASIC development is available. Wafer yields are much lower for photonic ICs, however, because waveguides require significantly more space than electrical paths. With silica-on silicon technology, the (silica) waveguides are formed using either chemical-vapor deposition or flame hydrolysis. The use of a silicon substrate allows passive and active devices to be combined. Another advantage is that silica waveguides also couple well to silica fibers because of their refractive index match. One drawback with this technology is the inability to realize large

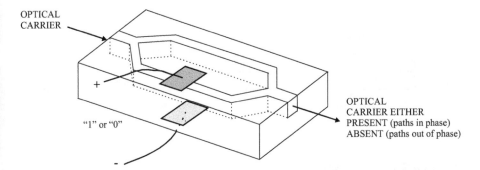

Figure 8-10   *Lithium-niobate modulator.*

refractive index differences for the "cores" and "cladding" of the waveguide. This leads to larger waveguide bends, which results in lower wafer yields.

A particularly attractive substance for substrates is *indium-phosphide (InP)*. InP provides the base material for long wavelength, III-V sources using InGaAsP heterostructures. Using InP, it is theoretically possible to monolithically integrate long wavelength lasers, modulators, WDM filters, isolators, amplifiers, photodiodes, and electronic circuits. Unfortunately, InP is not easy to use as a base material. Wafers are small and delicate, which leads to higher costs.

When the signal processing functions placed on an IC are exclusively optical, the package is referred to as an *optical integrated circuit (OIC)*. If the functions use electronic circuits, the package is an *opto-electronic integrated circuit (OEIC)*. A distinction can also be made between active and passive OICs. Using the historic definitions, if a device requires the addition of (non-signal) power to achieve its intended purpose it is active. By this definition the stand-alone lithium-niobate modulator in Figure 8-11 is an active device.

Three basic types of circuit construction are used to fabricate OIC/OEIC circuits or subsystems: discrete component, hybrid and monolithic. An OIC meant to achieve multiple functions but made with discrete optical components has relatively high assembly costs, high optical insertion loss, and generally lower speed. The individual components need to be spliced. An OEIC made with discrete components will have bandwidth limitations because its electrical connections are relatively long. In this regard, a useful approximation is that the connection length in any electrical circuit should be less than $\lambda/10$ to be considered a *lumped* rather than a *distributed* component. If the connection has to be considered distributed, transmission line echos can be a major problem. Based on this rule-of-thumb, a

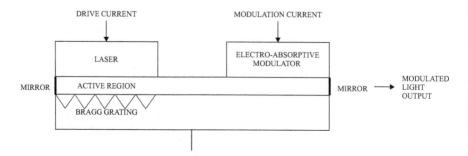

Figure 8-11 *InP-EML optical integrated circuit (OIC).*

1 mm connection on an OEIC would pose a problem for frequencies beyond 30 GHz. Since OIC/OEICs are intended primarily for high-speed applications, discrete component construction is generally avoided.

A leading example of commercially available, hybrid OICs is a high-channel count, silica-on-silicon WDM filter that uses phased-array-waveguide-gratings (PAWG). In integrated form, PAWG filters are roughly about 3 cm^2 in area. A 6-in. wafer yields about 7, 40 channel PAWGs. Sometimes *variable optical attenuators* (*VOAs*) are included on the substrate to provide the capability of balancing individual channel power levels. The VOAs are thermo-optically controlled through a separate electrical circuit. Achieving these same functions using cascaded thin film or Bragg filters and individual attenuators would result in higher costs, higher optical losses, and a larger footprint. Micro-optic gratings were initially used to obtain precise WDM filtering but their costs are high when compared to later technologies. Essentially the same results can now be achieved with etched (echelle) gratings in the substrates. The wafer yield is better than that obtained with PAWGs because the echelle gratings are smaller.

Monolithic devices are made from the two semiconductor families, GaAs/GaAlAs and InP/InGaAsP. GaAs substrates are used for the 850 nm window and InP substrates for the longer wavelengths. At the receiver, detectors have been integrated with preamps, postamps and TDM demultiplexers. Soon optical components on the front-end such as WDM filters will be added. At the transmitter, lasers have been integrated with modulators, isolators, and power monitors. A generalized cross-section of a monolithic, InP/InGaAsP *electroabsorptive modulated laser* (*EML*) is shown in Figure 8-11. These have been available commercially for a number of years in 14-pin butterfly packages for use in high bit-rate transmitters. The DFB laser is constantly in CW mode. The modulator either absorbs or transmits the carrier energy depending on the input binary signal. A tunable EML would have separate Bragg grating and gain sections. From Section 8.5.3, tuning would be achieved through modifying the temperature of the Bragg grating. The high-speed, GaAs signal drive circuitry cannot be grown on the same chip but is closely coupled in the same package. Integration of the drive circuitry can double the range of an EML transmitter to about 80 km at 10 Gb/s.

A promising area of research for hybrid integration uses polymer substrates. An example of these materials is silicon oxynitride. Polymers provide the necessary high-index difference needed for sharp bends in PLCs. The space required for a thermally tuned VOA can be reduced by a factor of about 100 relative to silica VOAs. Also, less power is needed for control.

8.5.8 *Wavelength Converters*

The optical add/drop (OAD) combination sketched in Figure 8-7 exemplifies the optical flexibility that is becoming more widely used and important in the national communications network. Going a step beyond, wavelength converters are used to translate an incoming signal from one wavelength to another. A typical application might be when a customer desires a 10 Gb/s Ethernet connection from point A to point C, but the same wavelength cannot be assigned through an intermediate node B because it is being used from B to C. With a converter at B, the signal on one wavelength from A to B can be changed to a second wavelength for B to C. If this is achieved optically, the A to C link is said to be transparent.

The simplest converters use an O/E/O translation. The output laser's center wavelength might be fixed or tunable. The electrical pulse stream is not regenerated, so degradations add through this type of converter. Another drawback is lack of transparency. Because the electrical interface is optimized for the bit-rate, an increase in the bit-rate necessitates a change in converters.

All-optical converters can handle any bit-rate because they do not contain an intermediate electrical path. One type of optical converter uses a laser purposefully operated in a gain saturation mode. An incoming signal at one wavelength is injected into the laser, and causes saturation of a second, lasing wavelength when a "1" is present. The new wavelength's power decreases, which results in a mirror-image output of the input signal- but at the different wavelength. This effect is referred to as *cross-gain modulation (XGM)*. Drawbacks of this approach are that the laser may add noise, and the input power level has to be high.

Other optical converters are being developed that make use of non-linearities that can occur in fibers, lithium-niobate waveguides, and semiconductor optical amplifiers (SOA). SOAs in the InP/InGaAsP material system are of particular interest because they operate in the long wavelength range. One of the reasons EDFAs became dominant as line amplifiers, instead of SOAs, was because of better linearity. As often happens in a fast paced technology, what was once a drawback is now a useful characteristic. In SOAs, three methods of generating new wavelengths can be used: the XGM of the laser, four-wave mixing (FWM), and cross-phase modulation (XPM). Generally significant pump energy is needed to activate the non-linear behaviors. An advantage of these approaches is their high S/N and speed of response. Finally, R & D also proceeds on optical regeneration.

8.6   Optical Networking

A telecommunications *network* can be most simply defined as a grouping of communications nodes that share some common interest. Often the network will have multiple connections or links between the nodes. This definition is purposefully imprecise because of the widely variable concepts of networking in the telecommunications environment. Networking is carried out at many levels within the national network, for many different purposes.

Networking exists at a number of levels, but not yet in any significant way at the optical level. For example, an IP signal currently will proceed from the west to the east coast through a number of different IP routers. That would be viewed as part of a network by the ISPs. The optical fiber SONET/SDH (TDM) systems used for the long distance part of this connection could be one of many in a cascaded series of WDM systems. The point-to-point digital link modeled in Chapter 7 would be only one part of that network.

The initial, and still the main use, for optical fiber communication is to provide point-to-point, high bit rate connections between networking nodes. This is the optical level or, in protocol terms, the physical level. At the nodes, networking is achieved once O/E conversion has taken place. The data is decoded, switched, and routed. Network intelligence and control resides in the higher-level SONET/SDH, ATM, frame relay, and IP equipment. By an earlier definition, this type of optical layer is opaque. With the advent of networking at the optical level, the optical layer will evolve from opaque to transparent and from "dumb" to "smart", however. The main advantages of these two trends will be to reduce costs, allow new services and grades of service, allow fast service provisioning, and improve overall reliability.

Networks using only optical connections route wavelengths or λs without O/E conversion. The technologies that will enable this are: DWDM, tunable dispersion compensators, tunable lasers, tunable filters, and optical cross-connects (OXCs). The tunable components should be controllable from a remote location for the network to be completely flexible.

DWDM is needed in an optical network because many channels have to be available. The tunable lasers can have their center wavelengths changed for restoration or service provisioning. Wavelengths can be dropped/added/redirected using the tunable filters and OXCs. If a failure occurs, optical cross-connects would be used to set up an alternate path through the optical level network. Tunable dispersion compensation is needed because accumulated dispersion will change in a channel if the path length is changed.

Due mainly to the growth of the Internet, national and international networks are rapidly becoming data centered (data-centric). As a percentage of traffic on national networks, data accounts for greater than 50% of the total and is growing faster than voice. At the peak of the Internet/dot.com bubble of the late 1990's, IP traffic was growing by at least 100%/yr. By 2003, growth had slowed to a "mild" 50%/yr. Small wonder then that the telecommunications pleisiosynchronous network (PSH) infrastructure needed improvement.

The Synchronous Optical Network (SONET) currently satisfies the need for efficiently melding voice and data services, but it is still not an optimum solution from the data communicator's viewpoint. Networking at the optical level is being looked at as one way to address this problem. Optical networking now appears particularly attractive for metropolitan networks (MAN) and metropolitan access. This is mainly because one of the key technologies, DWDM, is more cost effective when total fiber lengths are limited. In other words, shorter DWDM systems can be installed with lower cbk.

New data protocols will also be important in the development of optical networks. The data being transported will be able to use the optical layer's control and administration functions to offer QoS differentiation and improved reliability. Also, new protocols will make voice-over-Internet better.

8.7   Coherent Optical Systems

With the intensity modulation approach currently used exclusively in telecommunications fiber systems, the optical detector responds only to the received optical power. A detector/receiver that could respond to other characteristics of the incoming electromagnetic (EM) wave would offer much more capability. With radio transmission, for example, the receiver can respond to variations in amplitude, phase, or frequency, not just power. This type of signal processing is referred to as coherent because the received EM wave has a consistent, stable, and predictable phase front. With coherent optical transmission, all of the functions and advantages associated with conventional radio systems can be obtained at optical frequencies: low noise, efficient bandwidth usage, tuning over a broad range, electronic dispersion equalization.

Even though DWDM and external intensity modulation have proven their efficiency up to the present, coherent optical transmission remains a desirable target. Theoretical studies show a potential improvement in receiver sensitivity of 15–20 dB, depending on the baseband signal and type of modulation. Comparing bandwidth efficiencies, a 10 Gb/s intensity modulated signal using one channel in a 50 GHz DWDM system has a usage of 0.2 b/Hz. Microwave radio systems should have at least 1 b/Hz. The total bandwidth available from 1250–1650 nm is 50 THz. If coherent transmission could be used, the information capacity on a single fiber would be enormous.

A coherent optical system will probably use either homodyne or heterodyne detection. The latter is generally used in radio systems. A homodyne receiver requires a local oscillator (LO) frequency tuned to the incoming carrier frequency. Demodulation occurs directly in the photodetector. A heterodyne receiver has an LO offset from the incoming carrier by a predetermined amount called the intermediate frequency (IF). The IF is then demodulated. The LO is tuned to the incoming carrier +/– the IF. Optical coherent systems will probably use heterodyne detection (Figure 8-12). In an optical heterodyne receiver the IF will be in the microwave range because the signal bandwidth will be large. A generalized block diagram of a coherent optical system is shown in Figure 8-12.

The reason the detection works is because both the carrier and the LO frequencies are stable and have very little noise modulation. The carrier and LO wavefronts that mix in the combiner must have spatial coherence and precise frequency control. This requires lasers with very narrow linewidths. Theoretical studies show that PSK and DPSK modulation would need sources with linewidths of about 0.2% of the signaling bit-rate. This translates to 20 MHz for a 10 Gb/s signal. ASK and FSK modulation appears to be more tolerant.

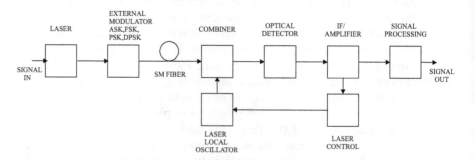

Figure 8-12 *Coherent optical transmission.*

Current high quality semiconductor sources used for intensity modulated systems have linewidths in the GHz range. Optical sources intended for use in a coherent system need about two orders of magnitude better stability. Diode lasers have a stability problem that might limit their use in coherent systems. Recall that for DWDM tightly packed channels, the laser can have a linewidth of only a few GHz. Achieving this is expensive, and requires careful package design and frequency control circuitry.

Appendix

Review Problems

CHAPTER 1—PROBLEMS

1. A commercial AM broadcast signal at 600 kHz has a wavelength of _____. A commercial FM broadcast signal at 1 MHz has a wavelength of _____. A communications carrier using a 6 GHz frequency has a wavelength of _____. All of these are through the atmosphere. Do you think the media will have an effect on the transmission?

2. The visible spectrum uses wavelengths between 400 and 600 nm. Compare this amount of spectral width to that used in the commercial AM radio band (assume 500–1700 kHz) in terms of the ability to convey "bits" of information. Assume an efficiency of 1 bit/Hz in both cases.

3. A power of 5 pW is equal to how many nW? A power of + 5 dBm is equal to how many mW? A frequency of 0.5 THz is equal to how many MHz?

4. How much slower or faster will a signal at 1.3 micron wavelength move through a GaAs laser than through a silica fiber?

5. A shorthand method that expresses a photon's energy uses electron-volts (EV). An electron volt is the energy needed to move the charge of an electron through a potential difference of I volt. The applicable relationship: Energy in EV = 1.24/(wavelength in microns). Derive the relationship between a photon's energy in EV to that in Joules.

6. A power sensitive light detector receives 1 microwatt. If the optical signal is centered at 850 nm how many photons/sec are received ? If centered at 1550 nm how many photons/sec are received ? If the carrier is modulated at 1 Gb/s, how many photons/bit are detected at the two wavelengths ?

7. An optical source launches +5 dBm into a 5 mile long fiber that has attenuation of 2 dB/km. What is the power level at the receiver in absolute level and dBm ?

8. The radius of the core of an optical fiber (a) relative to the free-space wavelength of the launched energy (λ) is a fundamental relationship that determines a fiber's performance. Compare a multimode fiber's 25 micron core radius at a wavelength of 850 nm to a single-mode fiber's 5 micron core radius at a wavelength of 1300 nm in terms of the total number of wavelengths. What do you think might be advantages and disadvantages to using these two different fiber types ?

CHAPTER 2—PROBLEMS

1. A wave of frequency 20 THz in one dielectric (refractive index = 1.48) strikes an interface with a second dielectric (refractive index = 1.46) at an incident angle of 30 degrees. What is the reflected angle? If there is refracted energy, what is its angle relative to the normal? What is the wavelength in the first dielectric? What is the wavelength in the second dielectric?

2. Almost all crystalline materials are anisotropic. They will have two possible values of phase velocity for a given direction of EM polarization. This means an EM wave, launched into the crystal with linear polarity, will have mutually orthogonal E-field components that are affected differently. What will be the state of polarization at the output and why?

3. A step index MM fiber is propagating an 850 nm signal in the steady state. The fiber has the characteristics listed in Table 2-1. What is the fiber's normalized frequency? How many modes will be propagating? Estimate the fraction of the total propagating energy that is carried in the cladding.

4. Using the fiber characteristics in Table 2-1, what is the critical angle of the plastic, step index MM fiber at the color red? How many modes will propagate in the steady state in this fiber?

5. A 6 ft. tall person beside a 25 cm deep creek spots a very small gold nugget in the water. It reflects light uniformly in a cylinder of one cm diameter. How close does he have to be, horizontally from the normal, to see the nugget? If the nugget reflects light in a cone of solid angle 10 degrees, how close must the person be to see it? Thinking he knows where it is he reaches directly for it. What happens?

6. A step-index fiber has a normalized frequency of 3. Estimate the number of propagating modes using Figure 2-11.

7. In a single-mode, step index fiber, the spot size (w) and core radius (a) are related by the empirical equation $w/a = 0.65 + 1.69V^{-3/2} + 2.8V^{-6}$. Plot the relationship of $I/I_{max} = e^{-2r^2/w^2}$ (Eqn. 2-16) for a = 5 microns, r = 1, 2, 3, 4 microns, V = 1,2,3.

8. Consider two single mode, step-index fibers with V=1 and V=2. Compare the amounts of power carried in the core and cladding. Which of these might be more susceptible to bending loss ? Why ? Which of these might be more susceptible to higher order modal noise? Why?

9. **(SPECIAL DESIGN PROBLEM)** A Gaussian beam from a 1300 nm vertical cavity surface emitting laser is to be focused onto a SM fiber (characteristics given in Table 2-1). The laser emits a cylindrical beam with a spot size = 50 microns. The projected Gaussian spot size at the focal point is given by the equation: spot size $w_0 = (\lambda \cdot$ Image focal length of lens$)/(\pi \cdot$ object spot size). Assuming a thin lens, design the spacing between the source-lens-fiber so that the projected spot size fills the fiber's core. Find the magnification using: Magnification = Image distance / objectdistance = Image size / Object size. What is the energy coupled to the fiber as a % of the laser's maximum, on-axis intensity, assuming no intervening losses in the lens.

CHAPTER 3—PROBLEMS

1. A ray directly down the axis of a step-index MM fiber (Table 2-1) will strike the core-cladding interface at the critical angle if the fiber is bent sharply by macrobends. What is that radius of curvature ?

2. In a step-index fiber, power is reflected for waves striking the core-cladding interface at greater than the critical angle ϕ_c. The EM field is continuous through the boundary in an "evanescent" field, which carries no power away from the core. The field decays as $e^{-x(2\pi\sqrt{(n_1^2 \sin^2\phi_1 - n_2^2)})/\lambda}$, where x is the distance along the normal from the

boundary, and ϕ_1 is the incident angle in the core. Using the step-index MM fiber characteristics in Table 2-1, an incident angle of 80 degrees, and a wavelength of 850 nm, what is the decay in field strength from the boundary after 1 micron penetration ? If the wave is at the critical angle, what is the penetration?

3. A fiber's 3 dB optical bandwidth is 1 GHz. What is the 3 dB electrical or baseband bandwidth ? Assuming a first order roll-off response, estimate the rise time attributable to the fiber.

4. Estimate the amount of modal dispersion in 5 km of the two silica MM fibers in Table 2-1 at a wavelength of 850 nm. Estimate the modal dispersion in 5 km of the plastic fiber. Do these results agree with the per-km bandwidth estimates given in the Table?

5. There is an "equilibrium" distance at which the modes in a MM fiber mix causing a decrease in the effects of modal dispersion. Up to this distance the modal time spread is linearly dependant on distance. Beyond this point it is dependent on the square root according to: $\Delta\tau_{mod-E} = \sqrt{L \cdot L_E}$ ($\Delta\tau_{mod}/L$), where L_E = equilibrium length. The $\Delta\tau_{mod}/L$ values are obtained using equations 3-8, 9. Assume the two MM silica fibers in Table 2-1 have equilibrium distances of 1 km. At 850 nm what are the modal spreads after 5 km? How do these results compare with those in Pr. 3-4?

6. Estimate the chromatic dispersion in 5 km of the two silica MM fibers in Table 2-1. Assume an 850 nm LED source with 50 nm linewidth. How do these results compare to the modal dispersion (Pr 3-5)? Next assume a laser with 2 nm linewidth, and recalculate the chromatic dispersion. Compare the two cases.

7. With standard SM fiber, the chromatic dispersion curve can be approximated in the region 1200 nm to 1600 by:

$$D_{chr.} = \left(\frac{D_0}{4}\right)\left(\lambda - \left(\frac{\lambda_0^4}{\lambda^3}\right)\right), \text{ where } D_0 = -0.095 \text{ ps/nm}^2 \cdot \text{km and } \lambda_0 = 1300 \text{ nm}$$

If a laser source centered at 1500 nm drifts +/– 1%, how much will be the variation in dispersion over a 100 km length of fiber ?

8. **(SPECIAL DESIGN PROBLEM)** A four wavelength, 1000 km SM optical fiber system uses wide-band optical amplifiers of 30 dB maximum gain, + 5dBm total maximum output, and – 30 dBm total minimum input. The optical amplifiers can be used at the transmitter, at intermediate locations, or at the detector as an optical preamp. The fiber plus necessary optical components has an average attenuation of 0.5 dB/km. Each wavelength's laser transmitter outputs 0 dBm and

each wavelength's detector must see a minimum of –35 dBm to maintain a proper S/N ratio. Design the system so that it uses a minimum number of amplifiers and meets all performance objectives.

CHAPTER 4—PROBLEMS

1. Using equation 4.1, show why the linewidth of an LED is wider at longer wavelengths given the same bandgap energy.

2. An LED is made of an alloy compound, $Ga_{1-x}Al_xAs$, where x = 0.06. The band gap energy in EV relates to the molecular structure using the following equation: $E_g = 1.424 + 1.266x + 0.266x^2$. What will be the wavelength of the LED?

3. Carrier lifetime, not parasitic capacitance, generally determines the high frequency, 3 dB bandwidth of an LED. The optical output power as a function of frequency is given by equation 4.3. If the average carrier lifetime is 1 ns, what is the LED's 3 dB modulation bandwidth? What is the optical 3 dB bandwidth?

4. A resistor, R, in series with a capacitor, C, connected to ground forms a first order filter when the output voltage is taken from across the capacitor. If the input voltage is $V_{in} = V_{max} \cos\omega t$ with $\omega = 2\pi f$ a variable, what is the 3 dB frequency ? What is the 10–90% rise time response to a step function? Show that the 3 dB frequency = 0.35/(rise time).

5. Above threshold, a laser outputs 1 mW CW when the drive current is 40 mA, and 3 mW CW when the drive is 45 mA. Plot the Power-out vs. current-in characteristic. Assume it is linear. Estimate the threshold current. Sketch the power output if a square wave drives the laser with an average current of 45 mA and max and min currents are 50/40 mA respectively.

6. The threshold current of laser diodes is dependent on temperature according to the empirical equation:

   $I_{th} = I_o e^{(T/T_o)}$ where I_o and T_o are constants

   With an InGaAsP laser at 1300 nm (I_o = 30 mA and T_o = 333° C), what will be the threshold current at T=150 degrees C? If a temperature control circuit keeps the operating temperature within 1 %, what will be the threshold variation ?

7. A DFB digital laser transmitter at 1550 nm uses an external, Mach-Zehnder, $LiNbO_3$, interferometric modulator. To achieve a 180 degree

phase shift in one of the two legs the required length must be at least 2 mm. The laser must keep coherence for at least that length. What must be the laser's time? What type of semiconductor laser would be a good choice?

8. The shape of the gain curve in a GaAs laser is given by a Gaussian shape:

$G(\lambda) = G(\lambda_0)e^{-((\lambda-\lambda_0)^2/2\alpha^2)}$ m^{-1} where λ_0 = center wavelength, α = curve width

Assume α = 50nm and the laser is 300 nm long. Assume also that stimulated emission is sustained when the gain is greater than ½ the peak value. This is called the FWHM bandwidth. How wide is the FWHM gain region? How many longitudinal modes will be in the output?

CHAPTER 5—PROBLEMS

1. An incident 1.55 micron, 2 microwatt optical signal from a fiber is coupled through an air-interface to an InGaAs photodiode with area larger than the fiber core. Assuming 100% of the energy is collected, what is the output current ? The responsivity plots in Fig. 5-5 already include the effects of the air interface. What is the conversion efficiency of the diode alone ?

2. The internal capacitance of a photodiode can be approximated by:

$C = (A \cdot \varepsilon_s/w)$ Farads, where A = area,

ε_s = permittivity = $1.03(10^{-10})$ F/m in Si, w = depletion width in m.

A PIN detector for an 850 nm, 50 micron diameter single-mode fiber has a depletion width of 25 microns. To collect all the light, the diameter of the photodiode is 1000 microns. If the load resistor is 50 ohms, estimate what the receiver bandwidth is if it is limited by RC rise-time alone.

3. Increasing the depletion width in a photodiode increases the carrier transit times. This will increase the bandwidth if the RC bandwidth is sufficiently high. A compromise is arrived at when:

w = depletion width = $(2/\alpha)$, where α = absorption coefficient in 1/m

The Si absorption coefficient at 850 nm = $8(10^4)$. Estimate the depletion width. Estimate the average transit time by assuming only the drift (depletion region) speed of electrons.

4. Using the bandwidth-to-rise-time relationship in equation 5.14, combine the two responses in problems 5–2, 3 to get a total rise time estimate for the Si detector. They can be combined on a RSS basis.

5. The average number of electron-hole pairs $\overline{N}$ generated in a photodiode in a time interval τ by an optical signal is determined using equation 5-4:

   $\overline{N} = (\eta \cdot E/h \cdot f)$ where η = detector quantum efficiency,

   E = Average Energy in τ

   The actual number fluctuates according to a Poisson distribution:

   $P(n)$ = Prob. of n electrons in $t = \overline{N}^n \, (e^{-\overline{N}}/n!)$

   If a BER of 10^{-9} is required, this must be the probability that a "0" is detected when a "1" was sent. Setting the probability to 10^{-9} and $n = 0$, the average number of E-H pairs per interval τ is found to be 20.7. If the quantum efficiency is 1, the photon limit becomes 20.7 in the interval. Assuming equal probability of a "1" or "0", the average received power at this quantum limit is given by equation 5.6. Figure 5-7 graphs this quantum limit at 1550 nm and efficiency = 100%. Find the quantum limit at 1300 nm and 100 Mb/s. How would equation 5.6 change if the required error rate was 10^{-11}?

6. A photodiode has characteristic curves as plotted in Fig. 5-6. Assuming the I-V scales are linear, estimate the maximum received power before saturation if the dc power supplied to the diode is 5 V and the load R=10 Kohms. If 10 microwatts is needed to meet the S/N objective, what is the dynamic range? If the load resistor is changed to 5 Kohms what is the result?

7. The average power of an 850 nm, 155 Mb/s signal into a Si detector is 2 microwatts. The load resistor is 50 ohms, the noise bandwidth is equal to the bit rate, and the preamp noise figure equals 4. Calculate the resulting S/N ratio at the preamp output.

8. Assume the preamp in Prob.5-7 has input C_A = 3pF, and R_A = 1mΩ. If the photodiode has C_D = 3pF, what is the possible circuit induced bandwidth limitation? What is the resulting risetime?

CHAPTER 6—PROBLEMS

1. Two GMM fibers with the characteristics listed in Table 2–1 are connected. The connector introduces 1 dB of loss. If the loss is assumed entirely due to lateral offset, what is the absolute amount of offset?

2. Factor K_1 in equation 1 for total SM fiber connection loss accounts for the reflection loss caused by the change of refractive indices between the transmitting fiber's silica core, an air interface, and the receiving fiber's silica core. Show that this loss is given by equation 6.6.

3. A physical contact (PC) SM connector for use with conventional SM (1300 nm) fiber has a lateral offset of 1 micron and an angular tilt of 1 degree. Using RSS addition, estimate the total loss.

4. A fused, biconical SM directional coupler has a splitting ratio of 4 to 1. In Figure 6-8A, approximately 4/5 of the power exits port B and 1/5 port D. If the coupler has an excess loss of 1 dB and the input at A is 10 mW, how much exits B & D?

5. Construct a star coupler matrix that uses cascaded 3 dB couplers that have 1 dB excess loss. There are eight inputs that accommodate eight different wavelengths, thus four input couplers. The couplers are sufficiently broadband to each transmit the eight wavelengths. Each of the eight output ports must have each of the eight wavelengths. What is the insertion loss of the coupler?

6. Bragg gratings are very useful in micro-optic form, as a tuning resonator for lasers, or as fiber embedded filters. Equation 6.7 relates a Bragg filter's spacing to the wavelength in the fiber material. If the spacing is 523 nm and the index of refraction of the fiber is 1.48, what is the tuned wavelength in air?

7. A LiNbO$_3$ phase modulator, 6 mm long, has a "voltage-interaction" length product ($V_\pi \cdot L$) of 16 V-mm. This is the combined voltage and length required to produce a phase shift of π radians. An application of 4 volts would shift the signal in 1 mm, $\pi/4$ radians. How long would a digital "on-off" modulator with the same characteristic have to be if the applied voltage is 8 V.

8. A Lambertian emitter is at ½ the peak value at what angle from the normal? The coupling efficiency into a GMM fiber with radius "a" > or = to the radius of the source is given by: Coup. eff. = $(NA)^2(1 - (1/2)(r_s/a)^2)$ where r_s = LED radius. If a fiber with the characteristics in Table 2-1 is used and the source has radius = 50 microns, what is the coupling efficiency?

9. **(SPECIAL DESIGN PROBLEM)** A connector has to be designed for GMM fibers to have a connection loss, excluding any reflection losses, of 1 dB or less. Your design must take account the four remaining

external factors and the two applicable internal factors in Table 6-1. Using your judgment, prioritize their importance and then assign their individual dB contributions to the total loss objective. Assume that their non-dB attenuation factors add on an RSS basis. Finally, translate these individual loss values into tolerances for your physical design.

Glossary

Chapter 1 — From RF to Optical Fibers

Media — The substance through which electromagnetic waves are propagated. Through the-air radio or optical waves use the atmosphere as a media.

Radio frequency (RF) — Electromagnetic waves in the frequency range from 10 kilohertz to 300 gigahertz.

Electromagnetic waves (EM) — Propagation of energy made possible by two field components, electric (E) and magnetic (H), acting in concert. Characterized by frequency, wavelength, velocity, and E-field polarization.

Photonics — technologies or components specifically developed to use wavelengths in the range of 1 micron (300 terahertz). Similar to use of electronics to describe electrical circuits.

Photon — A quantum or particle of light energy used to explain the ability of a light wave to displace electrons in a detector, thus creating a current. Characterized by energy and momentum.

Wavelength — The distance between repeating values of a propagating EM wave. The wavelength is equal to the velocity of propagation divided by the frequency.

Silica — The compound silicon dioxide (SiO_2). In purified form it is the most common ingredient used to make optical fibers.

Dispersion — Dependency of speed of propagation on frequency (wavelength). A signal with multiple spectral frequencies will suffer distortion because of dispersion because of the differing delays in arrival time.

Index of refraction — Describes the bending or refracting of light at the interface between two different substances. Equal to the ratio of the speed of propagation in a vacuum to that in the substance. Normally expressed relative to a vacuum, equals the square root of the product of relative permeability, μ_ρ, and permittivity, ε_ρ. In glass, μ_ρ equals 1.

Total internal reflection — Incident light on an interface will be totally guided if the angle of propagation relative to a normal at the surface is great enough.

Electrical to optical conversion (E/O) — Modulation of an optical source, a laser or an LED, by an electrical signal.

Optical to electrical conversion (O/E) — Detection of an optical signal at the front end of a receiver so that the electrical signal can be recovered.

Laser — A narrow spectra, semiconductor light source that can emit relatively high power, focused output. "Laser" stands for light amplification by the stimulated emission of radiation.

LED — A broad spectra, semiconductor light source with a wide angle output pattern. "LED" stands for light emitting diode. The light is generated by a spontaneous change in electron energy levels.

Digital Signal Processing (DSP) — The conversion of one signal, usually in analog format, to digital format and its subsequent processing while in the digital format.

Wavelength Division Multiplexing (WDM) — The combining of many optical signals to allow efficient transmission on a single fiber. Spacing between channels can be very close (Dense WDM) or relatively far apart (Coarse WDM).

Chapter 2 — Optical Fiber Characteristics

Ray optics — A geometrical approach to studying optical waves in optical fibers. Rays are used to describe the direction of propagation of the incident, reflected, and refracted components of a wave.

Angle of incidence — When a ray strikes an interface between two different materials, this is the angle between the direction of the ray and a normal to the surface.

Angle of reflection — When a ray strikes an interface between two different materials, this is the angle between the normal to the surface and the reflected ray. It equals the angle of incidence.

Step-index fiber — A fiber with an abrupt change in refractive index at the core-cladding interface.

Mode — A propagating pattern of electric and magnetic fields in a (fiber) waveguide caused by incident and reflecting waves alternately canceling and reinforcing each other. Pattern depends on wavelength, core size and shape, and material.

Mode groups - Different modes that travel at approximately the same velocity are referred to as a mode group. The three first higher order modes constitute a mode group.

Multimode fiber (MM) — A large core fiber that sustains multiple modes. Some propagate in the core and some that die out in the cladding.

Single-mode fiber (SM) — A small core fiber that sustains only the fundamental mode.

Graded index (GRIN)- In a multimode fiber, a core refractive index shape that approximates a parabola. Rays at the outside edge are bent to the middle, which leads to lower modal dispersion. Graded indexes are also used in lenses.

Graded index multimode fiber (GMM) — A multimode fiber in which the change between core and cladding indices is shaped gradual. Generally the shape is parabolic.

Modal dispersion — The multiple modes in a multimode fiber arrive at a detector at different times, thus leading to dispersion or spreading of the signal energy.

Material dispersion — The velocity of propagation in glass is dependent on the wavelengths carried by a signal, which leads to dispersion or spreading of the signal.

Waveguide dispersion — The shape of the core/cladding index of refraction will make some wavelengths travel at different velocities, which leads to dispersion or spreading of the signal energy.

Chromatic dispersion — The combination of material and waveguide dispersion.

Zero dispersion wavelength — Wavelength where chromatic dispersion is zero. Material dispersion will have a natural zero wavelength around 1300 nm. The addition of the smaller waveguide dispersion component to the larger material component makes control of this wavelength possible.

Plane wave front — The field values in an EM wave will appear constant in phase and amplitude, thus it is referred to also as a constant phase front.. The spherical shape to a radiating wave is considered flat over a small receiving aperture relatively far from the source.

Radiation pattern — The pattern of field intensity in a plane normal to the axial direction of radiation from a source.

Snell's law — Relates the direction of the reflected and refracted waves when an incident wave impinges an interface. The ratio of the two refractive indices equals the inverse ratio of their incident angles.

Critical angle — Ray angle, relative to a normal at the core/cladding interface, which results in total internal reflection. An incident angle less than critical results in energy being lost in the cladding and at bends.

Fractional index of refraction — The ratio of the difference between core and cladding indices divided by either the core or cladding index. Common values are in the region of 1 % or less.

Weak guiding — When core and cladding indices are very close, energy is also guided in the cladding. Only dielectric waveguides like optical fibers support this propagation.

Reflectance (R) — The ratio, 0 to 1, of incident energy to reflected energy at an interface. A 1 value indicates total reflection. Does not give values of any phase shift.

Polarization — The direction of the E field component of an EM wave. The H, or magnetic, component is orthogonal. In linear polarization, the E field is fixed in orientation. The E field rotates in circular and elliptical polarization.

Perpendicular polarization (S) — Relative to reflections, E field oriented 90 degrees to an (incident) plane determined by direction of propagation and normal to surface.

Parallel polarization (P) — E field oriented 90 degrees to the perpendicular polarization, in the plane of incidence.

Fresnel reflection laws — Give values of reflectance for S and P polarizations for at an interface for different values of incident angle and refractive indices.

Brewster angle — Incidence angle that results in no reflection at an interface. Differs for S and P polarizations.

Numerical aperture — Fractional amount of light that can be used by a device, like a fiber, processing light energy. Depends on size of aperture and internal indices of refraction. In fibers, varies from about 0.2 to 0.5.

Meridional rays — Rays in fibers that travel in planes that are cross-sectional to the fiber and contain the core's axis. They describe the transverse electric and magnetic propagating modes.

Skew rays — Rays in fibers that travel at angles relative to cross-sectional planes that include the core's axis. Skew rays follow an angular helix path as they propagate, and describe hybrid modes. Skew rays contribute significantly to leaky modes.

Leaky modes — Modes described by rays close to the critical angle (close to cutoff) will have energy propagating in the cladding, where it is more easily lost. Also called cladding modes. Non-coherent sources with broad radiation patterns (LEDs) will initiate leaky modes.

Propagation constant — Describes mathematically the amount of phase change an EM wave or mode/mode group experiences as it propagates in a medium. In a vacuum it is equal to 2π divided by free space wavelength λ. In a fiber, the component in the axial direction equals $2\pi N\sin\phi$ divided by λ. N is the effective refractive index of the wave or mode/mode group.

Phase velocity — Velocity of point of constant phase on an EM wave equal to ω divided by the propagation constant. Wavelength, λ, equals this velocity divided by frequency. In a vacuum, phase velocity equals the speed of light, c. A ray's phase velocity in a fiber equals c divided by the effective index of refraction. In a fiber, the phase velocity component in an axial direction equals c divided by $N\sin\phi$. Axial phase velocity will be greater than or equal to the ray's phase velocity.

Group velocity — Velocity of energy propagation of a wave, mode, or mode group. Equals the inverse of the derivative of the phase constant with respect to angular frequency. In a vacuum or an unguided medium, phase velocity equals group velocity. In a fiber, the group velocity component in the axial direction equals $\chi\sin\phi$ divided by the effective index of refraction, N. Axial group velocity will be less than ???? or equal to a ray's phase velocity.

TE and TM modes — Modal patterns in which the electric (E) or magnetic (M) fields are transverse to the direction of propagation down the fiber.

HE and EH modes — Hybrid modal patterns in which the field components are longitudinal in the fiber.

Cutoff wavelength — Wavelength above which a particular mode or mode group will not propagate. Depends on core radius and the fiber's numerical aperture. A SM fiber with radius 4 microns and a 0.1 NA would become multimode below about 1200 nm.

Normalized frequency (V) — A parameter used in step-index fiber analysis that combines four variables (core radius and index, wavelength, cladding index) to allow a broad view of modal performance. A fiber becomes SM if V<2.405.

Mode field radius — A measure of the spreading of the fundamental mode. In a step-index fiber, the shape is approximated by a Gaussian function. The mode field is more complex in SM fibers that have refractive index shapes optimized to control dispersion.

Spot size — Assuming a Gaussian shape for the fundamental mode, the radius is called the spot size and defined by the width where the amplitude is down $1/e^2$ or 13.5 % from the peak, on-axis value.

Matched cladding (MC) fibers — The standard SM fiber design. The glass external to the core has a uniform index of refraction, usually that of the starting glass, silica. The core is doped, generally with germanium, to have the required higher index.

Depressed cladding (DC) fiber — A SM fiber designed to have less attenuation, a smaller mode field, and less bending sensitivity. The cladding has an index lowered through Flourine doping. The zero dispersion wavelength for MC and DC fibers is around 1300 nm.

Dispersion shifted (DS) fiber — A SM fiber designed to shift the chromatic dispersion zero wavelength higher by controlling the waveguide dispersion component. The core is smaller and the fractional index difference is increased.

Dispersion flattened (DF) fiber — A SM fiber with a small core radius and annular rings. The zero dispersion wavelength is shifted upwards and the chromatic dispersion curve flattened.

Preform — Cylindrical rod that is a large version, radially, of desired fiber. Obtained through vapor-phase deposition or crucible melting. About a meter in length.

Fiber drawing tower — Heats the end of a preform and draws a fiber. Fiber is coated with thin protective film.

Tight buffering — A fiber with a thick protective layer added before cabling. One of two ways to protect a fiber from bending, chemical/water incursion, or damage. Also used to protect jumper fibers.

Loose buffering — A fiber with limited protective layering. It is placed in a tube to gain protection. The fiber has more freedom to move.

Dielectric cable — A fiber cable that excludes all metal. Used where lighting or ground loops from power systems might be a concern.

Chapter 3 — Optical Fiber Performance

Attenuation — Loss of signal energy, usually with respect to distance. Expressed in decibels, which are derived from 10 times the log of the ratio of two powers. Components in the signal path also have losses sometimes expressed in decibels.

Delay distortion — A signal is distorted when its optical wavelengths arrive at the receiver suffering differential time delays. Modal and chromatic dispersion in fibers are the main cause of delay distortion. Components will introduce bandwidth limitations that have a similar effect as delay distortion.

Linewidth — The width at the half power points of the spectral line generated by an optical source.

Dispersion — See Chapter 1, #8.

Polarization — See Chapter 2, #36

Linear system — A system or sub-system is linear if two signals, input simultaneously, do not interact. Further, no harmonics or sub-harmonics of input sinusoids should appear in the output.

Steady State — In multimode fibers, after a sufficiently long length (> meter), the modal structure becomes stable and remains so if no disturbances are encountered. Cladding and skew modes will not be present. This equilibrium state is not the same as occurs over kilometer distances when modes mix resulting in a decrease of the per-km effects of modal dispersion.

Infrared — Wavelength range above the visual red wavelength of about 600 nm and up to microwave wavelengths. Silica optical fibers are often referred to as operating in the near-infrared.

Ultraviolet - Wavelength range below the visual violet wavelength of about 400 nm, extending down to the X-ray region.

Resonance (molecular) — At certain wavelengths, substances can experience internal responses at the molecular level. An example in silica fibers is the absorption of certain wavelengths caused by the presence of water ions.

Rayleigh scattering — An attenuation phenomenon caused by the presence of minute variations in the refractive index. The small amount of lost light sets a lower limit on total fiber attenuation.

Optical time domain reflectometer (OTDR) — A test instrument that transmits an optical pulse and measures the time delay and amplitude of reflections. It is also used to measure losses in cables and components.

Macrobends — Relatively large-scale fiber bends caused by spooling and installation. Resulting losses generally are not significant, but fiber stress and damage a possibility.

Microbends — Small-scale fiber bends imposed during the manufacturing and cabling steps. Generally lower with loose tube cabling.

Full-width-half-max — For analysis purposes, describes the width of a pulse. In the time domain, it is the width of a pulse between points where the amplitude is ½ of the center (max) value. In the frequency domain, it is the width of the spectra between frequencies where the power is down 3 dB from the center (max) value.

Modal dispersion — See Chapter 2, #24

Material dispersion — See Chapter 2, #25

Waveguide dispersion — See Chapter 2, #26

Chromatic dispersion — See Chapter 2, #27

Time-of-arrival-distortion ($\Delta\tau$) — Amount of time spread added to a pulse as a result of arrival time distortion. In an ideal medium, all wavelengths in the source spectra would arrive at the detector without added time spread.

Root mean square (RMS) — Mathematical approach to calculating pulse spread in the time domain. Determined by taking the square root of the mean (average) of a variable over a defined time period.

First-order response — Describes a circuit frequency response that has a gradual roll-off of 20 dB/decade. The term "first-order" derives from the differential equations that describe a response in the time domain. Often associated with RC combinations.

Non-return-to-zero (NRZ) — A pulse format in which the pulse uses the whole allotted time slot.

Solitons — Pulses that are transmitted at wavelengths above the fiber's zero chromatic dispersion wavelength. If extremely narrow in time and of sufficient power, the fiber's dispersion can be compensated resulting in very long, undistorted transmission distances.

Polarization mode dispersion (PMD) — In single-mode circular fibers, the two orthogonally polarized components suffer differential time delays during transmission.

Birefringence — The degree of differential transmission between two orthogonally polarized waves. A high birefringence fiber could be used to isolate and protect one polarization. A zero birefringent fiber treats both polarizations equally.

Linear polarization maintaining fibers (LPMF) — Fibers constructed specifically to isolate and maintain a single propagating polarization.

Side-tunnel fiber — An LPMF designed to be only single-mode by suppressing any energy in the orthogonal mode.

Plastic optical fibers (POF) — Optical fibers, used primarily in the visual range, made generally of polymethyl methacrylate (PMMA). They use a step index and have large cores to increase NA. Used for short distances because of high attenuation.

Chapter 4 — Optical Sources for Fibers

Light emitting diodes (LED) — See Chapter 1, #14

Laser — See Chapter 2, #13

Continuous wave (CW) — A signal without modulation. In RF, this would be a single frequency sinusoid. With lasers and LEDs, the signal is a constant power output with multiple wavelength (frequency) components.

Bohr condition — In an optical source, the mathematical relationship between an electron's change in energy level and the frequency of the resulting optical emission.

Spontaneous emission — A photon of wavelength χ/ϕ is naturally emitted when electrons fall between levels that satisfy the Bohr condition. This is the only type of emission from LEDs.

Spatial coherence — The phasing is constant over a broad front of an EM wavefront. LEDs have spatially noncoherent wavefronts.

Temporal coherence — The phasing is predictable from one time to another at one position on an EM wave's propagation path.

Polarization coherence — The orientation of a linearly polarized wave, or degree and direction of rotation of a circular/elliptical polarized wave, is constant.

Recombination — When a free electron moves from a higher to a lower state and occupies a hole. The released energy generates a photon.

Direct-band-gap — In an optical semiconductor source using GaAs, InP, and InAs, recombinations between electrons and holes occur relatively easily because they both have the same momentum.

Indirect-band-gap — Does not allow direct recombination of electrons and holes. The electron first has to be forced to give up momentum. Silicon and Germanium are in this class and are not good optical emitters.

Lattice constant — Describes the strength of the molecular bonds in a crystal. The layered semiconductors used to make optical sources require lattice matching to resist internal stresses and strains.

Heterojunction — A boundary between layers in semiconductors. "Hetero-" refers to the use of substances with different atoms.

Homojunction — A boundary between two areas of the same substance, doped with donors and acceptor atoms. A pn junction is the best example.

Epitaxial growth — Technique used to fabricate semiconductors using gases to deposit successive layers.

Surface-emitting LED (SLED) — Also called a Burrus LED. The active, recombination region, is circular and radiation takes place vertically relative to the active layer.

Edge-emitting LED (ELED) — Similar to conventional semiconductor laser construction. The active, recombination region, is rectangular and radiation takes place from the edge of the device.

Superluminescent diode (SLD) — An optical source constructed like an ELED but with internal gain. Lasing is blocked because of a lack of mirrors giving a resonant cavity.

Lambertian emitter — A cone shaped radiation pattern from SLED that decreases from the center max value according to the cosine of the offset angle.

Injection laser diode (ILD) — A semiconductor laser that is controlled and modulated by current injected into the cavity.

Small signal response — The transfer function of an optical device. A low-amplitude sinusoid is swept through the baseband and the roll-off of the device obtained.

Large signal response — The response of an optical device to a relatively large amplitude high speed pulse. Lack of a good small signal response will lead to pulse distortion, and the device may introduce degradations at turn-on/off.

Modulation bandwidth — Also called the baseband or electrical bandwidth. Defined by the 3 dB down frequency in the small signal response, measured after demodulation. Since the electrical baseband signal tracks the optical power, the modulation (power) bandwidth is narrower than the optical bandwidth.

Optical bandwidth — The frequency at which the output optical power is down 3 dB below its peak, central optical frequency value. The subsequent modulation output will be down 6 dB at this frequency.

Chirp- In ILDs, the cavity index of refraction can change with the injected current density, thus causing a broadening of the linewidth. This results in an impairment and is one of the reasons external modulators are used in very high speed systems.

Rise (fall) time — Usually taken as the time for a pulse to grow (decay) from a 10% value to a 90% value.

Saturation — Occurs when a device is driven at a high input and the output no longer tracks volt for volt, or mA for mW in the case of an optical source.

Bias — A dc voltage or current, independent of any added signal voltage, that positions a device's operating point on its input-output transfer characteristic.

Stimulated emission — In a laser, if the drive current is sufficiently high, a photon will cause, or stimulate, the generation of more photons of the same wavelength.

Thermoelectric (TE) cooler — Used mainly for lasers to control device temperature. Uses n and p type semiconductor devices whose temperature gradients are current controlled.

Longitudinal modes — Since an ILD is basically a resonant cavity, the length, material, and internal reflections determine the wavelengths of oscillation. These modes are generated along the length of the cavity. Each resonant wavelength is actually a band of wavelengths because of internal laser noise.

Transverse modes — If the cross-sectional dimensions of a laser's active region are relatively large, the radiated output might have multiple modes, thus degrading performance and leading to "kinks" in the output curve.

Fabry-Perot laser — A multiple longitudinal mode laser. Wavelengths of oscillation are determined by cavity length and the lasing material.

Single-longitudinal mode laser (SLM) - An ILD with a sharp resonance can have only one longitudinal mode in its output. The resonance "Q" is enhanced through added tuning, as in the distributed feedback (DFB) laser.

Gain-guided ILD — A type of ILD active region construction. Variation in current density across the active region confines and controls the area of stimulated emission.

Index-guided ILD — A type of ILD active region construction. The side regions of the active region have different indices of refraction, thus providing an internal waveguide.

Kink — A discontinuity in the characteristic curve of an ILD caused by the sudden appearance of transverse modes in the output.

Threshold current — The drive current in an ILD where stimulated emission begins. At drive levels below threshold the light is generated by spontaneous emission.

Rate equations — A set of equations describing the electron and photon density rate of changes in a laser's active region.

Distributed feedback laser (DFB) — A single-mode ILD with sharp cavity tuning obtained by adding corrugations to adjacent longitudinal layers at the heterojunctions.

Footprint — Applies to the size of an electronic or opto-electronic package.

Erbium-doped fiber amplifier (EDFA) — An optical amplifier realized using a length of silica fiber doped with the element erbium. A pumping signal of an appropriate 2nd wavelength causes energy to be transferred to a signal traversing the fiber.

Amplified spontaneous emission noise (ASE) — Noise added to the desired signal by an optical amplifier.

Semiconductor optical amplifier (SOA) — An optical amplifier designed like a laser, but with the stimulated emission suppressed.

Raman amplifier — An optical amplifier that uses the non-linear characteristics of a silica fiber.

Vertical cavity surface emitting lasers (VCSEL) — A laser constructed similarly to the SLED. It will have lower manufacturing costs than the longitudinal ILD.

Chapter 5 — Optical Detectors and Receivers

Thermal detectors — Transform optical energy into heat. Used mainly in optical sensor sensor systems.

Photon detectors — Transform photons into an electrical signal. Transformation can be achieved directly into electrons, as in a photoemissive tube, or indirectly into electron-hole (E-H) pairs, as in optical semiconductors.

Transceiver — An equipment combination that includes both a transmitter and a receiver.

Photoemission — When a photon strikes a metallic surface with the proper coating, a single electron can be separated from the surface to become part of a current.

Photoconduction - The ability of certain substances to increase their conductivity with an increase in incident light energy.

Photodiode — A diode with an internal region where photons are transformed into H-E pairs, which yields a current driving an external circuit.

Positive-intrinsic-negative photodiode (PIN) — A basic p-n photodiode with a large intermediate intrinsic (non-doped) region that facilitates photon transformations.

Avalanche photodiode (APD) — A photodiode with voltages applied that are high enough to cause the multiplication of electrons and holes and generate an internal current gain.

Drift velocity — Velocity of carriers as they move through the depletion region in a semiconductor. Depends on the electric field strength and its profile.

Depletion region — At the junction between two oppositely doped substances, the holes and electrons will combine and deplete carriers in a small region of. A potential barrier is generated.

Absorption depth — In a photodiode material, the depth of penetration of light will experience before conversion to E-H pairs. The deeper the depth, the more efficient is the device.

Diffusion velocity — In photodiodes, carriers generated outside the area of elevated electric fields will experience diffusion at a slower velocity than the drift velocity.

Band gap energy — A photon can cause the generation of an E-H pair if it has the appropriate energy. This is the reverse process to the generation of photons in optical sources (Chapter 4, #96).

Responsivity (R) — A measure of a detector's ability to transform light energy into electrical current. It is the ratio of current output to optical power input.

Signal to noise ratio (S/N) — Generally expressed in dB, the required ratio of signal to noise at the baseband output of the receiver that will maintain adequate system performance.

Dark current — The noise generated in a detector when there is no optical input.

Sensitivity — The minimum detectable optical power given a required level of S/N at the receiver's output.

Poisson distribution — The statistical description of the transformation process that converts photons to E-H pairs in a semiconductor detector.

Quantum limit — The lowest possible sensitivity of a detector, assuming no dark current or added noises. It is calculated as a function of bit rate using a Poisson distribution.

Index of modulation (m) — Indicates the degree of carrier modulation in an analog system. A 100% modulation index occurs when the carrier is taken to zero and up to twice its normal value by a baseband signal.

Front end — Refers to the combination of photodiode detector and preamp used in an optical receiver.

Thermal noise — Temperature dependent noise generated in a resistor caused by the random fluctuation of electrons.

Shot noise — Noise in a device caused by the random fluctuation of electrons.

Excess noise factor (M^x) — The multiplication factor for shot noise in an APD of gain M. The exponent, x, is a function of the APD material.

Transit time — An approximate maximum time for carriers to transit the region where most of the photon-to-E-H pair transformations take place.

High impedance preamp — A preamp design with a high load impedance load on the detector. Maximizes S/N but limits bandwidth, thus requiring equalization.

Low impedance preamp — A preamp design with a low load impedance on the detector. Maximizes bandwidth but reduces S/N because signal from detector is lower and thermal noise higher.

Transimpedance preamp (TIA) — A preamp design that incorporates an op-amp instead of a detector load impedance. Maximizes bandwidth which makes it preferred for high speed systems.

Chapter 6 — Optical Components

Passive devices — Do not require energy from a power supply or an added signal input.

Connector — Used to make temporary connections between fibers. Affixed manually by an installer, or in a factory.

Splices — Used to make permanent connections between fibers. Achieved either through fusion or mechanical splicing. Splices have lower loss than connectors.

Couplers/splitters — Used to combine or separate an optical signal.

Wavelength division multiplexers/demultiplexers (WDM) — Used to combine or separate different wavelengths.

Filters — Used to pass or reject single or multiple optical signals.

Circulators — Used to transfer optical signals internally in the device from one port to the next adjacent port.

Isolators — Used to allow transmission in one direction while blocking any return signal.

Attenuators — Used to lower or control the power level of an optical signal. They generally find use in receivers and during test procedures.

Switches — Used to route an optical signal.

Modulators — Used to place the baseband signal on an optical carrier externally to the optical source. This avoids the need to inject the electrical signal directly into the source.

Full fill — Refers to the modal condition in multimode fibers where all possible propagating modes, even those that will quickly die out, have been excited.

Uniform fill — Same as full fill.

Steady state — See Chap. 3, # 69.

Scribe and break — Description of procedure followed to obtain a clean end-face on a fiber prior to attaching a connector or splicing.

Lap and polish — Description of final procedure followed to obtain a satisfactory fiber end-face after a connector has been attached.

Physical contact connector (PC) — A connector specifically intended to allow the end-faces of the two fibers to touch. This design can significantly reduce reflections.

Graded index multimode fiber (GMM) — A fiber with a core index of refraction profile, usually parabolic, that follows a gradual change from the peak axial value to that of the cladding.

Numerical aperture — See Chapter 2, #41

Return loss — The ratio in dB of reflected power to incident power. This ratio is alsoidentified as the reflectance (Chapter 2, #35).

Profile alignment (PAL) — One of two basic, automatic, fusion splicing methods. The profiles of the two fibers are aligned before fusion by using orthogonal light sources to highlight the fiber cores.

Light injection and detection (LID) — One of two basic, automatic, fusion splicing methods. Light is injected ahead of the splice location and detected on the other side.

Micro-optic devices — Very small optical components used in larger assemblies. These are usually lenses, gratings, etc.

Directional coupler — A bi-directional coupler that can be used to either combine two signals or split one signal.

Splitter — A directional coupler that has multiple ports.

Time division multiplexing (TDM) — The combining of digital signals into a faster bit stream.

Dense wavelength division multiplexing (DWDM) — Densely packed frequency division multiplexing in the optical domain. Channels can be spaced as close together as 0.4 nm (50 GHz).

Figure-of-merit (FOM) — A measure of the in-band flatness to the degree of roll-off in a DWDM device. It is derived by dividing the bandwidth at the 25 dB down points by the 0.5 dB down bandwidth.

Coarse wavelength division multiplexing (CWDM) — A low cost alternative to DWDM that spaces channels farther apart at 20 nm.

Bragg grating — A filter that uses periodic reflection points. These are used in SLM lasers and filters, essentially to give a sharp resonance wavelength to an optical device.

Arrayed waveguide grating (AWG) — A DWDM device that uses multiple Bragg gratings to separate and combine channels in a single step, thus reducing insertion loss.

Faraday rotation — A phenomenon used in optical isolators. The polarized input is rotated 45 degrees in a fixed direction through the application of a magnetic field. Any reflections are blocked by the input polarizer.

Bit-error-rate (BER) — The average number of errors detected in a digital transmission compared to the basic transmission bit-rate.

Circuit switching — A switching technique that makes a connection between two paths that remains for the duration of the required time.

Optical-electrical-optical switching (OEO) — An optical signal switching technique that converts the optical signal to electrical format, switches the modulated signal, and then reconverts it to optical.

Optical-optical-optical switching (OOO) — An optical switching switching technique that switches at the optical level.

Optical cross-connect (OXC) — An optical level matrix that allows any one input port to be connected to any of the output ports.

Non-blocking — The ability to cross-connect any input to any output on a matrix regardless of whether the matrix is being used by other signals.

Micro-electromechanical systems (MEMS) — Extremely small optical components developed using semiconductor technology.

Chirp — See Chap.4. # 116

Mach-Zender modulator — A modulator based on a Mach-Zender interferometer. The phasing between two optical carrier paths are varied by the baseband signal which intensity modulates the carrier.

Chapter 7 — Optical Fiber Systems

Regenerator — A regenerator achieves three "R"s: reshaping, retiming, and retransmission of a digital signal. Currently this is done by an O/E transformation, followed by regeneration, and then modulating an optical source.

Baseband — The bandwidth of the electrical signal driving the optical transmitter. (see also Ch.4, 114)

Frequency division multiplexing (FDM) — In analog systems, the stacking of channels very close together to realize a broader baseband.

Time division multiplexing (TDM) — In digital systems, the interleaving of pulses from independent pulse streams into a faster pulse stream.

Synchronous optical network (SONET) — The original North American digital multiplexing standard.

Synchronous digital hierarchy (SDH) — The original non-North American digital multiplexing standard. SONET and SDH have common line bit rates at high speeds.

Synchronous transport module — Describes the ascending bit-rates of the electrical baseband signals used in the SDH standard. STM-1 is a rate of 155.52 Mb/s.

Synchronous transport signal (STS) — Describes the ascending bit-rates of the electrical baseband bit-rates used in the SONET standard. STS-3 is at 155.52 Mb/s.

Local area network (LAN) — A limited distance, data communications network installed generally on campuses or in buildings.

Metropolitan area network (MAN) — A large geographical area, data communications network generally installed using rings of optical fibers.

Asynchronous transfer mode (ATM) — A format/protocol system for digital transmission that uses frames made up of 5 header bytes and 48 data bytes.

User-to-network-interface (UNI) — In a packet switched network, a router would directly connect to an optical network. The optical network would provide open interconnections between IP and optical protocol layers.

Digital subscriber line (DSL) — A digital connection to homes and small businesses that provides bit rate service from a few hundred kb/s to 10 Mb/s.

Fiber distributed data interface (FDDI) — A standard for LANs that uses token ring and two directional ring transmission, thus providing a high level of reliability.

Bit-error-rate (BER) — See chapter 6, #198

Error function (erf) — The basic error curve in a digital system with S/N as the independent variable. It assumes white noise and equally probable "1"s and "0"s.

Line codes — A bit stream is coded before transmission to optimize performance. Each media might require a different type of line coding.

Non-return to zero (NRZ) coding — One of two basic line codes used for optical fiber systems. A pulse or zero uses the whole time slot allotted for one bit of information..

Return-to-zero (RZ) coding — One of two basic line codes used for optical fiber systems. A pulse uses only part of the time slot allotted for one bit of information.

Bit-rate — The rate at which bits of information are transferred in a digital system. A binary line codes (NRZ, RZ) will have one bit of information per time slot.

Baud rate — With data, the number of signal transitions per second. In binary systems, the baud rate equals the bit rate. With multilevel coding, the baud rate will be lower.

Symbol rate — The speed of transmission of symbols.

Baseline wander — With NRZ coding, a long series of "1"s will result in a significant amount of low frequency energy. Because of dc blocking, this results in changes in the baseline reference used to detect the presence of pulses.

Eye diagram — Obtained when many detected pulses are overlaid on an oscilloscope screen. Decisions about pulse presence or absence in the center of a time slot are made with little chance of error if the eye is open.

Forward error correction (FEC) — A type of digital coding that adds redundancy to the bit stream to detect and correct for errors made during transmission. This is added to the baseband before pulses are modified for NRZ or RZ modulation of the optical source.

Automatic repeat request (ARQ) — A type of error correction that requires both directions of transmission. If errors are discovered, the data is repeated upon request.

Automatic gain control (AGC) — A feedback system that reduces or increases the gain in a receiver if the incoming signal is high or low, respectively.

Intersymbol interference (ISI) — Because communication channels are always bandwidth limited, pulses will experience a rounding of their rise/fall transitions. This causes interference with neighboring pulses.

Relative intensity noise (RIN) — Noise generated by the random fluctuations of a laser's output intensity.

Reflection noise — Noise induced in a laser's output when reflected energy interferes with the cavity oscillations.

Mode partition noise (MPN) — Noise induced in a laser's output by the transitory appearance of multiple longitudinal modes. Occurs generally when pulsed on and off.

Side-mode suppression ratio — Describes the ratio in dB of the power in the main mode to the next highest side-mode power. Generally used only for single-mode lasers.

Chirp — See Chapter 4, #116

Modal noise — Noise that occurs with multimode fibers when a relatively coherent laser source is used. It is different than modal distortion.

Amplified spontaneous emission (ASE) noise — A noise generated in optical amplifiers caused by the spontaneous recombination of electrons and holes.

Polarization mode dispersion — See Chapter 3, #87

Polarization dependent loss — An optical signal attenuation caused by unequal response in a component to the incoming state of polarization. Most fiber systems cannot sustain a linear state of polarization.

Raman scattering (SRS) — One of a five phenomena caused by fiber nonlinearities. Optical power is transferred from shorter wavelengths to longer wavelengths.

Stimulated Brillouin scattering (SBS) — One of five phenomena caused by fiber non-linearities. In a SM fiber, the optical signal interacts with acoustic waves that occur randomly. An interference is scattered back that is offset lower in wavelength.

Self-phase modulation (SPM) - One of five phenomena caused by fiber non-linearities. An optical signal suffers phase modulation because the fiber's index of refraction (Kerr effect) is slightly dependent on signal intensity. Used to realize soliton transmission.

Cross-phase modulation (XPM) — One of five phenomena caused by fiber non-linearities. Similar to SPM, it occurs between channels in WDM systems.

Four-wave mixing (FWM) — One of five phenomena caused by fiber non-linearities. A problem with DWDM. Three wavelengths can mix and generate a fourth, interference, wavelength.

Nonzero dispersion shifted fiber (NZDSF) — The chromatic dispersion zero wavelength is shifted out of the (DWDM) band to cause a small amount of in-band dispersion. This acts to minimize FWM.

Interchannel crosstalk (ITXT) — Interference into the signal channel from another channel in a WDM system.

Intrachannel crosstalk (IAXT) — Interference into the signal channel from the signal itself.

Timing jitter — The "noise" on the extracted timing signal at a receiver that is used to judge when to decide on the presence or absence of a pulse.

Extinction ratio — In an optical source, the ratio of the output power in the on or "1" state to that in the off or "0" state. Lasers are always going to have some output with a "0" input because they are never taken completely past the threshold to zero.

Solitons — Very short pulses with controlled pulse width and energy. By using the intensity dependency of the Kerr effect they can be transmitted long distances.

Link power budget — The attenuations of all components in a link from transmitter output to detector input are analyzed and budgeted based on the receiver sensitivity requirement.

Link rise time budget — The rise times, hence bandwidth performance, of all components in a link are analyzed and budgeted based on the receiver S/N requirement.

Reach — The length of fiber plus components between regenerator locations.

Optical add-drop (OAD) — Through the use of filters and WDM equipment, optical signals can be added to a fiber at dropped at intermediate locations on a link.

Optical loss test set (OLTS) — A test instrument that combines optical sources and detectors to measure power and arttenustion.

Optical time domain reflectometer (OTDR) — see Chapter 3, #69

Optical back-reflectometer — A test instrument used to measure reflections

Optical spectrum analyzer (OSA) — A test instrument used to measure the spectra of optical signals and the bandwidths of optical components.

Multiwavelength meter (MWM) — A test instrument used to measure wavelengths.

Differential group delay (DGD) — The average differential delay between the two polarizations in a single mode fiber that leads to PMD.

Chapter 8 — Future Directions-Telecommunications

Dot.com — A term referring to the rapid growth in Internet technology and use in the late 1990's. In this context it includes the frenzied growth in optical fiber placement intended to meet the needs of the "dot.com" growth.

lit — A term applied to a fiber that is transmitting an optical signal, with or w/o modulation.

Reach — see Chapter 7, #257

Long haul system — A system with a reach of about 600 km.

Ultra-long haul system - A system, usually using submarine cable, with a reach in the thousands of kilometers.

Metro system — A system with a reach of about 100 km.

Legacy fiber — Optical fiber to be used for a new system that was previously installed. It may not have all the required characteristics for the intended new system installation.

Cost-bit-km (cbk) — A measure of the overall efficiency of a fiber system. Includes the cost of placement and operations.

Voice-centric network — Refers to the usage of the National/International communications network from a voice or telephony viewpoint. Circuit switching is used.

Data-centric network — Refers to the usage of the National/International communications network from a data viewpoint. Packet (Internet) switching is used.

Transparent optical network — A network that allows optical connections between end points. Implies the use of WDM, optical amplifiers, couplers, switches and higher-level control.

Opaque optical network — A network that requires EOE and OEO conversions between end points. Implies the use of regenerators, TDM drop-add, and low-level control.

Public switched telephone network (PSTN) — The circuit switched telephone network that uses TDM to carry many different voice and data signals. Private, dedicated lines can be leased.

Plesiochronous digital hierarchy (PDH) — The digital hierarchies in North America and Europe in use before SONET/SDH. The PDH used many independent clocks. The lower bit rates up to 45 Mb/s are still in use.

Server — The equipment that connects a customer to the Internet. Capable of a range of input and through-put speeds, and customer services.

Loop — The telephone network connection to the subscriber. Term applied to the need for an electrical path, or loop, to get the voice signal to the customer.

Broadband passive optical network (B-PON) — In a passive optical network (PON), connections to customers are achieved using directly connected optical fibers without active components like regenerators, repeaters or amplifiers. Broadband means that wideband connections are available.

Fiber-to-the-curb (FTTC) — A distribution network that makes optical fibers available at the curb. A customer would be connected electrically at that point.

Fiber-to-the-home (FTTH) — A distribution network that makes optical fibers available directly to the home.

Backbone — A path in a network which aggregates traffic in order to gain economy of scale advantages.

Bell System — A vertically integrated combination of companies headed by the American Telephone and Telegraph Company (AT&T) that was broken apart in 1984 by the US Government. Local operating companies provided direct services to customers. The Long Lines Division provided interconnections. The Western Electric Company provide product development and manufacture. The Bell Laboratories was the R&D branch.

Quality-of-service (QOS)- A specified level of service provided to a customer. Generally carries a bit error rate and outage time commitment.

Calling party — The person initiating a telephone call.

Trunk — A circuit switched path in the telephone network used for toll traffic.

Tandem trunking — Trunks between toll offices in the telephone network that cross to other toll offices without following the switching hierarchy.

Central office (CO) — The lowest office in the telephone switching hierarachy that provides service directly to subscribers. A loop originates in a CO.

Circuit connections — In the voice-centric network, a direct connection or path set up for the full period of a telephone call.

Virtual connections - In the data-centric network, a packet based connection that results in different parts (packets) of a message being sent over different physical paths.

Frame relay — An early packet switching protocol intended to link LANs and provide higher speed data transport (up to 2 Gb/s). Along with its predecessor, X-25, frame relay networks are "overlays" on the existing TDM transport systems.

Asynchronous transfer mode (ATM)- A telephone derived packet-based protocol/switch for the interchange of digitized voice, data and video. The top service grade allows circuit-switched, real-time, connections. Fixed cells of 53 bytes are used, with the first 5 identifying the data and destination.

Internet protocol (IP) — A digital data, packet-based protocol that uses variable length packets. The current widely used version (IPv4), does not allow priority assignments, hence has problems with time delays for some services. A new version (Ipv6) will allow priority assignments and thus will be better for real-time signals like voice.

Analog to digital conversion (A/D) - To facilitate TDM or packet transmission, an analog signal must first be coded into digital form. The A/D conversion of a voice signal results in a 64 kb/s pulse stream, for example.

Integrated services digital network (ISDN) — A early telephone network offering intended to provide end-to-end digital linkage for customers and a range of services. Originally offered two 64 kb/s voice lines and a 16 kb/s data line. Now called integrated digital subscriber line (IDSL) to provide service beyond the reach of DSL.

Digital subscriber line (DSL) — A family of digital connections to subscribers that provide speeds up to tens of Mb/s over existing wire pairs. Distances and speeds are limited by the quality of the wire pair loops.

Information revolution — A term generally applied to the rapid growth in data transmission and use that has occurred over the past two decades.

Core network — A general term used in the data-centric view that refers to the long-distance, widely interconnected parts of the national/international network.

Wide area network (WAN) — A general term used in the data-centric view that refers to a sub-network within the Core network.

Metropolitan area network (MAN) — A general term used in the data-centric view that refers to a high speed data network of a few hundred kilometers in an urban area. Often a MAN will link LANs and enterprise locations.

Access network — A general term used in the data-centric view that refers to the local connections to MANs, WANs and the core.

Edge — A general term used in the data-centric view that refers to the interface between a customer and the greater data network.

Packet-switched network — A way of routing information that breaks a signal into packets with a header that is used to guide the signal through the network. The multiple packets that constitute the signal may not all follow the same path.

Routers — The equipment in a packet network that reads the headers and chooses the path through the network. They operate mainly at the network-layer level, 3, of the OSI structure (312).

Protocols — The "rules" by which different layers in the data hierarchy exchange information.

Statistical multiplexing — Rather than using assigned spaces to interleave incoming bits, as in TDM, a stat-mux stores and then combines them according to traffic demands. More time is allocated to the busiest channels.

Internet service provider (ISP) — At the "edge" of the data-centric network, ISPs provide the required connections to local customers. Using servers, they generate the packetized signals that are transmitted through the Internet under the control of routers.

Datagram — An older, general term used to describe a self-contained data message.

International Standards Organization (ISO) — The international body that deliberates on, and publishes, standards relative to data transmission. Many national organizations, such as the IEEE in the US, have working committees to support the ISO.

Open systems interconnection model (OSI) - A 7-layer model used as a guideline for understanding and developing data communications systems and protocols. Hardware is included in the lower two levels, Physical and Data-link. Software defines the upper 5 ½ layers.

Physical, Data-link, Network layers — The lower three layers of the OSI that guide the design and use of optical fiber systems. As optical networking grows, more Data-link and Network functions will be performed within the optical network.

Overhead — The amount of capacity in a data communications connection that has to be assigned for administrative uses such as protection switching, error control, routing, etc.

Internet — A network of (data) networks that use packet switching. IP,s use variable-length packets. The protocol used in the Internet is the Internet Protocol (IP).

Transmission control protocol/Internet protocol (TCP/IP) — A layered set or grouping of protocols used to give access to the Internet for a variety of services.

Scalability — A general term that describes the ability of a network or equipment to grow or add capacity without a substantial investment.

Dry fiber — SM fiber that has no attenuation peak at 1308 nm, hence can be used through the range from 1200 to 1600 nm.

Zero dispersion shifted fiber (ZDS)— SM fiber that has a core/cladding design that places the dispersion zero in the middle of the longer wavelengths to be used.

Non-zero dispersion shifted fiber (NZDS) — SM fiber that has a shifted dispersion curve, but the zero is purposefully not at the wavelengths to be used. This fiber is used when non-linear effects, such as four-wave-mixing, need to be minimized.

Dispersion compensating fiber (DC) — SM fiber designed with a large amount of dispersion that is used to compensate for dispersion built up from longer lengths of installed fiber.

Dispersion profile — A plot of dispersion through the length of a SM fiber link that includes dispersion compensation.

Power profile — A plot of power levels through the length of a SM fiber link that includes all attenuation factors and optical amplifier gains.

Repeater — An opto-electronic amplification of a signal without regeneration. Repeaters are often used in local networks.

Erbium doped waveguide amplifier (EDWA) — An optical amplifier used in hybrid optical ICs. Embedded waveguides are used for the amplification medium instead of fiber.

Raman amplifiers — Optical amplifiers that use a high power counter-propagating pump energy to provide gain for the desired signal. They are used throughout the 1300 — 1550 nm range.

Semiconductor optical amplifier (SOA) — An amplifier that is basically a laser without the reflecting mirrors.

Hybrid amplifier — A combination of an Erbium doped fiber amplifier (EDFA) and a Raman amplifier that results in flat gain over a broad range of wavelengths.

Wavelength add-drop filter (WAD) — A wavelength sensitive filter that allows the addition or subtraction of a single wavelength, or group of wavelengths, at a network node. There are a number of technologies used to realize this function.

Zero water peak fiber (ZWPF) — The same as dry fiber (Chap.8, # 318).

Arrayed laser — An approach to realizing a tunable optical source. Lasers with different output wavelengths are arrayed on the same chip which allows the choice of the desired wavelength.

Wide-tuned lasers (WTL) — A packaged laser with a tunable range of about 100 nm.

Narrow-tuned lasers (NTL) — A packaged laser with a tunable range limited to a few nanometers.

Planar lightwave circuits (PLC) — A reference to optical ICs that use waveguides embedded in substrates.

Pockels effect — A linear electro-optic response used in PLC devices to achieve a variety of functions.

Indium phosphide — A basic crystalline material used to build long wavelength optical sources and optical ICs.

Optical integrated circuit (OIC) — An IC that contains only optically based components.

Opto-electronic integrated circuit (OEIC) — An IC that contains both optical and electronic components.

Lumped components — Usually referring to transmission lines, when the connection length is relatively short the transmission effects can be ignored and components can be treated as separate and distinct.

Distributed components — When transmission lines are relatively long, the effects of the connection must be taken into account using values assigned on a per-unit length basis.

Variable optical attenuators (VOA) — An optical power attenuator that can be adjusted. Often used on OICs performing a WDM function.

Electroabsorptive modulated laser (EML) — A type of external modulation that places the modulator on the same substrate as the laser.

Cross-gain modulation (XGM) — A wavelength converter that uses a saturated laser to develop a different output wavelength.

Bibliography

BOOKS:

Agrawal, G. and Dutta, N., *Long-Wavelength Semiconductor Lasers*, New York: Van Nostrand Reinhold, 1986.

Barnoski, M., *Fundamentals of Optical Fiber Communications*, New York: Academic Press, 1976.

Bates, R. and Gregory, D., *Voice and Data Communications Handbook*, New York: McGraw Hill, 1998.

Bell Telephone Labs, *Engineering and Operations in the Bell System*, Murray Hill, New Jersey, 1984.

Bell Telephone Labs, *Transmission Systems for Communications*, Murray Hill, New Jersey, 1982.

Daly, J., *Fiber Optics*, Boca Raton, Florida: CRC Press, 1984.

Diament, P., *Wave Transmission and Fiber Optics*, New York: Macmillan, 1990.

Freeman, R., *Fiber-Optic Systems for Telecommunications*, New York: John Wiley & Sons, 2002.

Geckeler, S., *Optical Fiber Transmission Systems*, Norwood, Mass.: Artech House, 1987.

Gowar, J., *Optical Communication Systems*, Englewood Cliffs, New Jersey: Prentice Hall, 1984.

Green, L., *Fiber Optic Communications*, Boca Raton, Florida: CRC Press, 1993.

Hecht, J., *Understanding Fiber Optics*, Upper Saddle River, New Jersey: Prentice Hall, 2002.

Jeunhomme, L., *Single-mode Fiber Optics*, New York: Marcel Dekker, 1983.

Jones, K., *Introduction to Optical Electronics*, New York: Harper and Row, 1987.

Jones, W., *Introduction to Optical Fiber Communication Systems*, Orlando, Florida: Holt, Rinehart and Winston, 1988.

Kadambi, J and Kalkunte, M., and Crayford, I, *Gigabit Ethernet, Migrating to High-Bandwidth LANs*, Upper Saddle River, New Jersey: Prentice Hall PTR, 1998.

Kao, C., *Optical Fiber Systems: Technology, Design, and Applications*, New York: McGraw Hill, 1982.

Kartalopoulos, S., *Introduction to DWDM Technology: Data in a Rainbow*, New York: IEEE Press, 2000.

Keiser, G., *Optical Fiber Communications*, New York: McGraw Hill Higher Education, 2000.

Leon-Garcia, A. and Widjaja, I., *Communication Networks — Fundamental Concepts & Key Architectures*, New York: McGraw Hill, 1999.

Lin, C. (Editor), *Optoelectronic Technology and Lightwave Communications Systems*, New York: Van Nostrand Reinhold, 1989.

Malik, Om,, *Broadbandits, Inside the $750 Billion Telecom Heist*, New York: John Wiley & Sons, 2003.

Miller, S. and Kaminow, I., (Editors), *Optical Fiber Telecommunications II*, San Diego, California: Academic Press, 1988.

Midwinter, J., *Optical Fibers for Transmission*, New York: John Wiley and Sons, 1979.

Mukherjee, B., *Optical Communications Networks*, New York: McGraw Hill, 1997.

Palais, J., *Fiber Optic Communications*, Upper Saddle River, New Jersey: Prentice Hall, 1998.

Personick, S., *Fiber Optics, Technology and Applications*, New York: Plenum Press, 1988.

Powers, J., *An Introduction to Fiber Optic Systems*, Chicago, Illinois: Richard D. Irwin, 1997.

Tomasi, W., *Electronic Communications Systems: Fundamentals through Advanced*, Englewood Cliffs, New Jersey: Prentice Hall, 1994.

Suematsu, Y., *Introduction to Optical Fiber Communications*, New York: John Wiley & Sons, 1982.

Wickersham, A., *Microwave and Fiber Optics Communications*, Englewood Cliffs, New Jersey: Prentice Hall, 1988.

Wilson, J. and Hawkes, J., *Optoelectronics: An Introduction,* Englewood Cliffs, New Jersey: Prentice Hall, 1983.

Yariv, A., *Optical Electronics*, Orlando, Florida: Holt, Rinehart and Winston, 1985 (also 1997).

PERIODICALS:

Bell Labs Technical Journal (www.lucent.com)
Electronics Letters (IEE)
Fiberoptic Product News (www.fpnmag.com)
IEEE Communications Magazine (www.comsoc.org)
IEEE Journal of Lightwave Technology
IEEE Journal on Selected Areas in Communications
IEEE Photonics Technology Letters
Laser FocusWorld (www.optoelectronics-world.com)
Optics & Photonics News (www.osa-opn.org)
Photonics Spectra (www.photonics.com
Photonics Tech Briefs (www.ptbmagazine.com)
Photonics Web Directory (www.photonics.com)
Private & Wireless Broadband (www.privatebroadband.com)

WEB SITES: (www.——)

Agilent.com (good tutorials)
Commspecial.com
Cooper.edu/engineering/projects/
Corning.com
Fiberopticsonline.com
Geocities.com/SiliconValley/Circuit/8070/ (good portal)
Internetphotonics.com
Lightreading.com
Lw.pennet.com
Nortel.com
Photonics.com
Photonics.cusat.edu (good portal)
Photonicsonline.com
Play-hookey.com/optics
Opticsnotes.com
SPIE.org
VPIsystems.com (design tools)

Printed in the USA
CPSIA information can be obtained
at www.ICGtesting.com
JSHW021319221024
72173JS00001B/7